SpringerBriefs in Architectural Design and Technology

Series Editor

Thomas Schröpfer, Architecture and Sustainable Design, Singapore University of Technology and Design, Singapore, Singapore

Indexed by SCOPUS

Understanding the complex relationship between design and technology is increasingly critical to the field of Architecture. The *Springer Briefs in Architectural Design and Technology* series provides accessible and comprehensive guides for all aspects of current architectural design relating to advances in technology including material science, material technology, structure and form, environmental strategies, building performance and energy, computer simulation and modeling, digital fabrication, and advanced building processes. The series features leading international experts from academia and practice who provide in-depth knowledge on all aspects of integrating architectural design with technical and environmental building solutions towards the challenges of a better world. Provocative and inspirational, each volume in the Series aims to stimulate theoretical and creative advances and question the outcome of technical innovations as well as the far-reaching social, cultural, and environmental challenges that present themselves to architectural design today. Each brief asks why things are as they are, traces the latest trends and provides penetrating, insightful and in-depth views of current topics of architectural design. *Springer Briefs in Architectural Design and Technology* provides must-have, cutting-edge content that becomes an essential reference for academics, practitioners, and students of Architecture worldwide.

More information about this series at http://www.springer.com/series/13482

Carlos BAÑÓN · Félix RASPALL

3D Printing Architecture

Workflows, Applications, and Trends

 Springer

Carlos BAÑÓN
Singapore University of Technology
and Design
Singapore, Singapore

Félix RASPALL
Universidad Adolfo Ibáñez (UAI)
Santiago, Chile

ISSN 2199-580X   ISSN 2199-5818 (electronic)
SpringerBriefs in Architectural Design and Technology
ISBN 978-981-15-8387-2   ISBN 978-981-15-8388-9   (eBook)
https://doi.org/10.1007/978-981-15-8388-9

This Springer imprint is published by the registered company Springer Nature Singapore Pte Ltd.
The registered company address is: 152 Beach Road, #21-01/04 Gateway East, Singapore 189721, Singapore

We dedicate this book to our families, who have supported us in the past five years. Starting up AirLab required much time, dedication, and support. We thank our friends from all over the world, who accompanied us in the process.

Foreword

3D Printing Architecture impressively brings together myriad important technologies that will play key roles in the application of 3D Printing to architecture. What sets it apart, and ahead of, the current literature is how it impressively integrates computational generative design of complex structures, simulation of functional performance, and multiple manufacturing process centered on 3D Printing into new digital workflow constructs. This results in a digital thread throughout the entire design and production supply chain including materials, manufacturing logistics, assembly logic, cost, and validation and verification. 3D Printing is the integrative production technology that ties these together, enables new advances in the individual technologies, and sets the stage for a digital design and production paradigm where increased design freedom is availed, informed design decisions can be made earlier in the process, and appealing system effects are exploited throughout the value chain.

Particularly, impressive is the array of case studies that have continually developed and adapted new technologies, integrated them to enable more creative and efficient design and production workflows, and demonstrated them in award-winning architectural products ranging from furniture to multi-functional installations to structural pavilions. These have taken 3D-printed architecture to new heights. While the examples highlight the power of the 3D Printing technologies when integrated in digital workflows focused on architecture, the implications and knowledge are much more broadly impactful to other types of products ranging from aerospace structures to medical devices. Researchers and practitioners in these and related fields will learn much from this book and it will shape their perspective on the future of product development.

A unifying thread through the case studies and workflows is how multiple manufacturing and assembly technologies are integrated in a way that maximizes the potential of each and does so early in the design phase. BAÑÓN and RASPALL cleverly exploit the unique capabilities of 3D Printing to advance a new framework for structural connections in architecture that are typically difficult to practically achieve for freeform geometries due to cost and/or mechanical integrity. Any structural connection has to both enable the continuity of the geometric form and provide the required mechanical and functional performance and durability. A major challenge, in addition to the assembly and fabrication ones, is the complexity of analyzing and qualifying the mechanical performance of the connection, either through existing design codes or more sophisticated but expensive numerical analysis and experiments. The 3D-printed node approach elegantly solves this by decoupling the geometrical continuity and mechanical integrity elements of a joint by creating a complex-shaped structural node that connects to (possibly many) standard structural members in any geometric configuration by using simple standard joining methods. The standard joints between the structural member and the node can be analyzed and qualified by established methods of stress and failure analysis. The body of the node then enables the complex geometric continuity. While it has a complex shape itself, the parametric design technologies BAÑÓN and RASPALL have developed allow it to be generated in an automated way that can control stress concentrations while imparting an appropriate aesthetic quality consistent with the overall architecture. While the shapes are complicated, 3D Printing allows their fabrication as a homogeneous solid which then can also be analyzed and qualified by methods of 3D stress and failure analysis routinely used in various engineering fields. Moreover, all of this is implemented in computational modeling and generative design tools that simultaneously determine the macroscopic structural architecture and the mesoscopic connecting node geometries. Within this digital thread, they have demonstrated the continual addition of increasing capabilities and sophistication throughout their case studies, building in logic for assembly, cost, and multi-functionality. The overall concept broadly suggests how in an automated way during design, one can decouple necessary manufacturing technologies and employ them, and specifically 3D Printing, where they can make the highest impact.

I think the perspectives that BAÑÓN and RASPALL have developed and articulated here arise from a rare commitment to highly collaborative and interdisciplinary work. I had the pleasure to see much of this firsthand at the dynamic new Singapore University of Technology and Design (SUTD) where they launched the highly innovative AirLab in 2015 and started their 3D Printing work. Since then they have attacked this topic in a unique way building teams of highly creative and skilled architects, designers, and engineers and integrating them with experts in materials, software, hardware, and systems. This highly collaborative team-based approach allowed them to engage at the system-level, as opposed to component-level typical in 3D Printing, and enable the impressive breadth of insights they provide. The culture they created in the AirLab, and how it contributed to and benefited from that of SUTD, was instrumental in how they rapidly moved the case studies presented here from concept to award-winning reality.

There is no doubt that BAÑÓN and RASPALL have illuminated an impressively broad array of issues required to successfully exploit the paradigm-changing capabilities of 3D Printing in real architecture. I think this will open the eyes of others, in architecture and other fields, about how important these issues are and attract increased attention to advancing them. While so much has been demonstrated here, we are just scratching the surface of realizing large-scale adoption of

3D Printing in practice. Much research and development remain to be done across all of the elements of the digital design and manufacturing workflows as well as with the workflows themselves. This spans multiple fields, including design algorithms and technologies, new 3D Printing materials and processes for faster, more consistent printing, failure analysis in complex 3D-printed geometries and materials, and optimization of supply chains for the manufacture and assembly of large-scale architecture. All of these, when integrated into seamless digital workflows, will unlock unprecedented innovation in mass customized architecture and product development. It is inevitable that this will happen. As it does, *3D Printing Architecture* will be recognized as a catalytic contributor to its happening.

May 2020

Martin L. Dunn
Dean and Professor
College of Engineering
Design and Computing
University of Colorado Denver
Denver, USA

Preface

Existing literature on additive manufacturing in architecture is limited to technical books for manufacturers or enthusiasts and projects, reflections, and examples of 3D Printing in architecture are abundant but scattered throughout online media. However, a book that compiles discussions on the implications of 3D Printing for architecture and associated design methods is still absent. This book aims to fill that gap, presenting discourses on creative and technical opportunities of 3D Printing. It includes a discussion of how this new technology relates to contemporary architectural trends, enables more creative and effective design workflows, and engages in a vast range of applications.

3D Printing enables new workflows, which incorporate dynamic computer models and address real-time cost control, material compatibility, structural properties, durability, and assembly sequence, and which directly interface 3D model to production. 3D Printing enables new applications, from ornamental panels to load-bearing, complex structural connectors. And ultimately, 3D Printing allows for new architectural trends, including computer-generated efficient forms, highly articulated freeform surfaces, and sustainable spaces.

Singapore
Santiago, Chile

Carlos BAÑÓN
Félix RASPALL

Acknowledgements

This book is dedicated to all the people and institutions that helped advance our research and innovation over five years.

Our gratitude goes to the outstanding team of designers, researchers, and assistants, who have been the soul of AirLab in the past and present: Anna Toh Hui Ping, Tay Jenn Chong, Aurelia Chan Hui En, Felix Amtsberg, Sourabh Maheshwari, Natalie Chen, Syahid Bin Mustapa, Wang Sihan, Nahaad Mohammed Vahid, Rajkumar Velu, Chadurvedi Venkatesan, Vedashree Jathar, Jonathan Ng, Mohit Arora, Michele Sodano, Hu Yuxin, Goh Yiping, Shiqian Pan, and many others.

The design research work that led to this book has been institutionally and financially supported by the Singapore University of Technology and Design (SUTD), the ZJU-SUTD IDEA Research Collaboration Programme, the Center for Digital Manufacturing and Design (DManD), the International Design Center (IDC), the SUTD Marketing Team, Lope Lab, and the National Design Center Singapore. The book would not have been possible without their generous support.

Contents

Chapter 1
Introduction

Abstract This chapter examines how architecture is shifting as designers start to adopt additive manufacturing as a powerful fabrication method in full-scale built designs. Current and emerging trends, such as a search for formal and spatial complexity, the use of ornament, and the search for sustainability, are discussed. In addition, changes in the design workflows, where information from the construction phase is readily and effortlessly available at early design stages, are reviewed. Key applications in the construction industry being explored at the moment are presented. Finally, the ten projects by AirLab are introduced as case studies to illustrate the potential of advanced manufacturing technologies for architecture.

Keywords Additive manufacturing · Computer-aided design · Design to fabrication · Design workflows · Digital architecture

Architectural design progresses in tandem with improvements in material technologies. Today, 3D Printing technologies deliver faster, bigger, stronger, and cheaper outputs. Novel workflows from the early concepts to subsequent project development, advanced manufacturing processes, and integration into fully functional products become available. The integration of parametric modeling with performance optimization in the design process is redefining the design process, as the material is computationally allocated where it is the most needed. As a result, forms are efficient and geometrically more intricate. Hence, 3D Printing becomes indispensable by enabling the manufacturing of these intricate artifacts where other well-established technologies fall short.

With 3D Printing, bespoke forms and multi-functional parts bear the same cost as standard ones; hence, complexity becomes inexpensive, and intricacy can appear at a very conceptual stage. An entirely new set of design opportunities, spanning from extreme levels of precision to minimal tolerances and seamless transitions between printed and standard non-printed components, requires the use of custom-built software and dynamic parametric models to conceptualize and advance a new conception of architecture.

Through a review of literature and a detailed study of ten built projects developed by the AirLab, this book reflects upon the disciplinary implications of an ongoing additive manufacturing revolution. A study of the main architectural trends behind the interest and adoption of 3D Printing technologies in architecture, as well as the main applications and design workflows, are presented through ten in-depth case studies of projects developed by AirLab over the last five years.

The selection of projects includes:

- *AirMesh*, a pavilion at the Gardens by the Bay, Singapore. 2017–2020
- *Timescapes*: Singapore University of Technology and Design's 10th Anniversary Time Capsule. 2018–2019
- *Sombra Verde*, a gazebo at Tanjong Pagar. 2017–2018
- *White Spaces*, the Singapore Pavilion at the Seoul Biennale. 2018
- *(ultra) Light Network*, an interactive structure at Marina Bay. 2016–2017
- *OH Platform*, a display table at the Venice Biennale of Architecture. 2018
- *Metadata*, a design exhibition at the National Design Center. 2018
- *Makerspace*, a design exhibition at the Urban Redevelopment Authority of Singapore. 2019
- *v-Mesh*, a gathering space and display platform at SUTD. 2015–2016
- *AirTable,* furniture design at Digital Manufacturing and Design Laboratory (DManD). 2017–2018

The projects represent a wide range of scales, programs, and applications, from one-off furniture pieces to complete exhibition designs and outdoor pavilions. Some of the projects are permanent architectural objects, while some had a brief duration of a few months. Some are conceived as static artifacts, and some are movable and reconfigurable objects.

As a group, these projects explore real examples of how additive manufacturing connects to and expands contemporary architectural trends such as freeform design, computational optimization, ornament, sustainability, and integration of systems. In addition, they elaborate design workflows, covering parametric modeling, optimization, file-to-factory manufacturing, and complex assembly sequence design. Individually, the projects engage on the specific topics of structural optimization, material selection and transition, mechanical properties, manufacturing process, logistics, and assembly procedures. Figure 1.1 exhibits a collection of built designs.

The ten designs utilize a range of printing technologies, tackling different applications, and dealing with several scales and cost constraints. Moreover, these components interact with standard industrialized members in singular strategies. Therefore, the technical design and the manufacturing process of each printed component constitute a reliable source of information regarding printing times, costs, mechanical performance, and design constraints. In most of the projects described in this book, the use of 3D Printing is limited to the fabrication of the structural knots, while the struts are typically fully standard or customized off-the-shelf profiles. In this way, the cost is reduced by combining both 3D Printing, which is still comparatively expensive, and affordable components

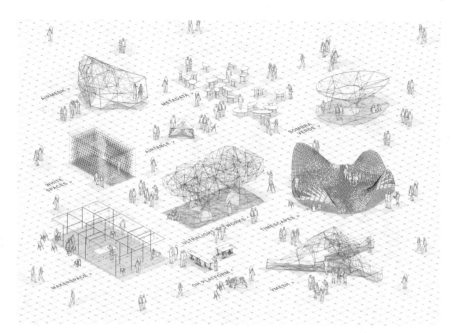

Fig. 1.1 Isometric drawings of the ten projects described in the book, on the same scale

available in the market. The most representative component for each project is shown in Fig. 1.2.

The projects were developed at AirLab, a design research laboratory co-founded by the authors of this book that focuses on design innovation through the application of digital technologies. Since 2015, a multi-disciplinary team constituted by architects, engineers, and designers, explores new creative processes at the intersection of architecture, digital design, advanced manufacturing and material science. The laboratory produces exploratory research through the development of digital tools and workflows and functional prototyping. It actively collaborates with the industry and research agencies, promoting the application and early adoption of its research in full-scale projects such as those discussed in this book. Through its body of research, AirLab incrementally refines and expands each step of the design-to-fabrication process and demonstrates results in real life.

1.1 Trends

Architectural design continuously evolves, driven by the creative search for innovation. Changing cultural conditions and technological possibilities inform and are informed by corresponding architectural trends. Today, a large portion of experimental architectural design is focused on understanding what digital culture can do

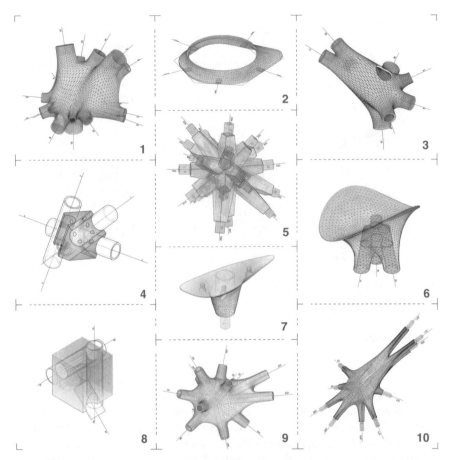

Fig. 1.2 Typical 3D-printed component for each of the ten case study projects. 1: *AirMesh*, 2: *Timescapes*, 3: *Sombra Verde*, 4: *White Spaces*, 5: *(ultra) Light Network*, 6: *OH Platform*, 7: *Metadata*, 8: *Makerspace*, 9: *v-Mesh*, 10: *AirTable*

for the built environment. The first digital revolution in architecture in the 1990s made computer-aided design mainstream, as an interest in formal complexity challenged the supremacy of simplicity and austerity. Mario Carpo compiles the positions, motivations, and experimentation of critical players in the digital turn in architecture in his edited book [1].

Several architectural trends evolved. Firstly, the simplicity in the creation of curvilinear, freeform surfaces, and highly articulated expressions through parametric design has inspired an expressionist trend. Digital technologies opened the way to architectures that can deal with curvature, fluidity, articulation, ornament and controlled variation and disorder. Secondly, computational structural and energy modeling started a performative trend. Designs address structurally optimal forms looking for lightness, controlled durability, differentiation based on functional parameters, and integration and applicable factors. Key architectural trends empowered by 3D Printing include expressionist and performative approaches.

1.1.1   Expressionist Trend: Formal Complexity

The development of parametric modeling software made the creation and precise control of complex geometries available to architects. Software originally tailored to computer graphics animations and car and aerospace design was applied to the design of physical spaces. From freeform NURBs surfaces to complex part assemblies, the parametric computer-aided design provided a platform to control increasingly complex projects. However, with an exponential increase in the number of unique parts, manufacturing and logistics of projects became progressively more challenging. In architecture and construction, computer numerical control fabrication (CNC), primarily CNC routing, became the go-to solution for complex geometry projects.

For the manufacturing of complex geometry projects, 3D Printing offers new realms of possibilities. Not limited by subtractive methods, where times and wastage can quickly become immense, additive manufacturing can accomplish true freeform projects with little formal constraints and reduced waste.

1.1.2   Expressionist Trend: Ornament

The digital revolution in architecture renewed attention to the ornamental dimensions of projects. As modern architecture broke up with ornament, the interest in and crafts of decorative elements became marginal. Recently, within the explorations in generative modeling, designers encounter the possibility of creating an infinite level of detail and articulation. This opportunity appears within the architectural culture of post-modernism, which had already reinstated ornament and decoration into the design scene. A thorough argumentation on the relationship between architectural ornaments and digital technologies is presented in Antoine Picon's Ornament: The politics of architecture and subjectivity [2].

In this context, 3D Printing has an edge over other manufacturing technologies such as CNC milling, because more articulation does not require necessarily more material or machine time.

1.1.3   Performative Trend: Structural Optimization

Structural modeling software, once accessible only to specialized engineers, is becoming more accessible and compatible with modeling tools familiar to designers. Therefore, the collaboration between design and engineering gets more fluid and dynamic. Kara and Bosia build this argument through the concept of design engineering [3]. The integration of structural modeling into parametric design modeling tools enables the development of generative models in which designs are automatically adjusted to enhance structural performance. Optimized structures are highly efficient, but often, at the expense of increased geometric complexity. As structural requirements continually change throughout a structure, the geometry and topology of optimized projects reflects these variations with highly differentiated and non-standard components. 3D Printing is an ideal technology for structurally optimized projects because it can achieve complex geometry and bespoke components without increase in cost and manufacturing effort.

1.1.4 Performative Trend: Sustainability Optimization

Measuring the environmental impact of manufacturing processes, such as their contribution to global warming, is commonly underestimated or even neglected in the construction sector. Most studies focus on the embodied carbon in materials including their transportation, but crucial factors such as the amount and type of materials used shall be taken into consideration. Recent advances in manufacturing technology offer great potential to reduce the carbon footprint of manufactured goods and mitigate their contribution to global warming. Additive manufacturing can contribute to lower the levels of energy utilized in the production, ranging from 4 to 21% [4], with the most substantial effects in the feedstock, conveyance, and usage phases.

The direct benefits of additive manufacturing in conjunction with data-driven digital models also contribute to other environmental issues, such as reduction of material consumption, wastage, water usage, and emissions. From this point of view, additive manufacturing can contribute to achieving a circular economy model. The development of parametric modeling tools with energy modeling functionalities brings environmental information to designers during their design stage and ultimately improves building's performance [5].

1.2 Workflows

Traditionally, the design process in architecture goes from conceptual sketches to a highly technical project, followed by the construction phase. Although each region names design stages differently, the usual steps in the process include five main phases: schematic design, design development, construction documents, tendering, and construction administration.

The design process, by definition, anticipates a future building, its qualities and use, structural stability, energy performance, budget, time, and the steps in the construction process. In order to achieve this, designers progressively develop a project from concept to technical solution. First, intentions and ideas are studied, which are then formed into a more defined geometry. This form is then progressively calibrated to accommodate technical, functional, and spatial requirements of the project. In the design development and construction documents phase, the project's technical definition increases as the project's structure, mechanical services, lighting, construction detailing, and other technical dimensions are assessed and specified by the designers and the consultant team.

The access to faster and more accurate technical information earlier in the design process can enhance the quality of a project. Therefore, digital modeling tools that can handle spatial and expressive dimensions of buildings (e.g., form, space, light, and texture) while handling their technical aspects (e.g., structural, construction, mechanical, budgeting and scheduling) within a single modeling environment maximize informed creativity. This argument is often presented by proponents of Building Information Modeling (BIM), stating that BIM can integrate technical information at an early design stage [6].

Additive manufacturing represents a breakthrough in the digitally driven workflow, as the designers can get almost immediately information about the manufacturability of building components, including cost, time, and mechanical specifications of parts. Besides, as 3D Printing replaces the use of standard parts, it is possible to envision components where crucial technical systems can be integrated into a single complex element. For example, fluid transportation, data, power, tags, IDs, fastening details can all be included in the printed component. Additionally, additive manufacturing simplifies the translation from design to construction, as the parts can be manufactured directly from the computer without the need to produce construction documents and written technical specifications. Figure 1.3 shows how digitally informed design using 3D Printing can enhance the design creativity and reduce uncertainties at the fabrication, assembly, and operation stages.

1.2.1 Design Phase

With additive manufacturing, the digital design process is enhanced through increased design freedom and level of control over the subsequent manufacturing, assembly, and building operation phases. With fewer constraints in the manufacturing of parts, designers have more geometric freedom to experiment in their projects. For aesthetic or performative reasons, architects can fully utilize the possibilities of computational design and envision complex structures that could not be achieved without additive manufacturing. In addition, the use of associative models adds to the geometric modeling of a quantitative depth. Designers are more aware of weights and mechanical properties when working with additive manufacturing (AM) as part of the workflow.

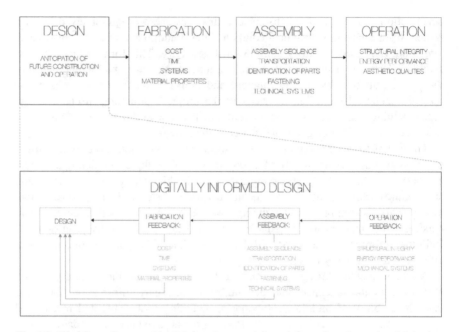

Fig. 1.3 Workflow diagram highlighting how real-time information from the fabrication, assembly, and operation stages feedbacks into the design

However, a whole new set of restrictions limits the application of existing additive manufacturing technologies in construction: reduced build volumes, long production times, high printing and equipment costs, limited selection of materials and finishes, and untested connection details, among others. Additionally, in contrast to traditional manufacturing methods, the lack of uniform norms and design recommendations constitutes a drawback for industry adoption. Designers' creativity comes to life when exploring additive manufacturing technologies, navigating the potentials of 3D Printing while working inside its technical possibilities.

1.2.2   Manufacturing Phase

The production of building components with additive manufacturing adds a new level of control and possibilities for the construction industry. Its capacity to determine with precision where materials are allocated increases the automation of an industry that is still heavily reliant on skilled labor. With more control over the process, risks to human life are reduced. Additionally, file-to-factory approaches, where digital models specify the form and composition of parts, the frequent human errors, and ambiguities in the interpretation of construction document disappear.

1.2.3   Assembly Phase

Through digital design and additive manufacturing, the complexity of projects can increase exponentially. The number of unique components grows, and the logistics of the construction site becomes daunting. The assembly process, therefore, must be planned accordingly. With unique and precise components adequately identified, the function and effectiveness of construction drawings are no longer the same. Cartesian coordinates are no longer necessary. They are replaced by computer models, spread sheets, and topological labeling between components.

Also, the bespoke sequence of construction becomes crucial to complete the assembly. Traditionally, buildings are separated into well-defined systems such as load-bearing structure, cladding, and mechanical systems. The construction process is typically clear as to what is built first and what follows. As 3D-printed components can perform many of these traditional functions simultaneously, a carefully planned step-by-step sequence shall be computationally verified before starting the assembly works.

1.3   Applications in Architecture

To date, the main application of 3D Printing in architecture has been the production of scale models for representation purposes. However, as printing technologies evolve, new applications for functional building components are emerging. Buildings are much larger than other objects typically produced by 3D Printing; therefore, only some parts are buildable with existing printing technologies.

Three architectural elements are starting to be produced with 3D Printing: ornamental panels, space frames, and concrete components.

1.3.1   Ornamental Panels

A substantial portion of the architectural applications are non-structural, decorative panels. This is a low-hanging fruit for 3D Printing. On the one hand, as they are non-structural, the mechanical requirements are less critical, allowing the use of polymers, the most widespread, and affordable printing option. Panels can also be printed with clay, providing a large range of glazing options and long durability. On the other hand, the focus on ornament engages with the fact that in 3D Printing, complexity in the object does not necessarily cost more: a simple and plain object will cost the same as a complex and ornate piece of similar dimensions.

1.3.2   Structural Connectors

Space frame applications are an ideal case for 3D Printing. Their geometric and structural complexity is concentrated in the small volumes of the nodal joints, where the largest proportion of the structure is achieved with standard components. Due to the reduced size of the nodes, the use of higher performance materials such as metals, composites, and technical polymers is feasible. The interface between printed and standard components, discussed in-depth in Chap. 6, is therefore crucial to ensure a compatible mechanical behavior of the parts.

1.3.3   Load-Bearing Walls

Concrete components constitute an important application of 3D Printing. This material is available at a low cost, its structural capacity is appropriate for buildings, and it is already established in the construction industry. Additionally, the processing of concrete is relatively simple, as it can be deposited in a fluid state and chemically transition to solid without additional processing steps, such as melting or binding. The most widespread concrete printing method is direct concrete extrusion, which is fast and straightforward, but limited in the range of geometries it can accomplish. Most of the designs are vertical walls, although some experiments have included more daring forms in columns and slabs, where concrete supports are painstakingly chiseled out. Alternatively, concrete components are produced with printed molds in plastic or clay, augmenting the design possibilities but requiring additional pre- and post-processing steps.

1.4   Organization of the Book

This chapter serves as an introduction, presenting the concepts and projects that guide the contents of the book. Chapter 2 reviews a selection of established and emerging additive manufacturing technologies and how they relate to the specific needs of architecture. Chapter 3 focuses on the optimization trend, looking at 3D Printing as an enabler of highly optimized structures, while Chap. 4 centers on the expressionist trend, examining the possibilities for ornamental and very articulated architectures. Chapter 5 looks at the potentials of 3D Printing for the use of sustainable material such as bamboo. Chapters 6–9 focus on aspects of the new design workflows: how freeform possibilities of printed parts can interface with standard components, how technical systems are integrated into multi-functional components, how real-time cost information can be used as a driver of design decisions, and how the assembly sequence for complex structures can be developed at an early design stage, respectively. The final chapter summarizes the conclusions from each chapter and presents an outlook for the potential of 3D Printing in architecture.

References

1. Carpo M (ed) (2013) The digital turn in architecture 1992–2012. Wiley, New York
2. Picon A (2014) Ornament: the politics of architecture and subjectivity. Wiley, New York
3. Kara H, Bosia D (2016) Design engineering refocused. Wiley, New York
4. Verhoef LA, Budde BW, Chockalingam C, Nodar BG, van Wijk AJ (2018) The effect of additive manufacturing on global energy demand: an assessment using a bottom-up approach. Energy Policy 112:349–360
5. Peters B, Peters T (2018) Computing the environment: digital design tools for simulation and visualisation of sustainable architecture. Wiley, New York
6. Goldman G, Zarzycki A (2014) Smart buildings/smart (er) designers: BIM and the creative design process. Building information modeling BIM in current and future practices, pp 3–16

Chapter 2
Additive Manufacturing Technologies for Architecture

Abstract This chapter covers the existing and emerging 3D Printing technologies and their potential and limitations for architectural design. Aspects related to the cost of production, size limitations, speed and mass production, material properties, and environmental benefits are discussed. A selection of the most relevant printing technologies for architecture is analyzed in more detail, including direct concrete printing, fused deposition modeling, direct metal laser sintering, and binder jetting.

Keywords Additive manufacturing · Fabrication methods · Post-processing · 3D Printing in architecture · Digital manufacturing

Additive manufacturing (AM), popularly known as 3D Printing, is the process of building objects bottom-up, one layer at a time. This approach to manufacturing provides two key advantages over previous methods: the geometric freedom in the components to print and the lack of specialized tools to produce the parts. In contrast, traditional mass manufacturing techniques build objects by subtracting or forming materials into the desired shape using tools and jigs, after which several items are assembled to form the final product.

In the 1980s, the world saw the invention of additive manufacturing with the first 3D Printing called stereolithography (SLA). The objective was to provide a cost-effective and time-efficient way to produce low-volume, customized products with complicated geometries for prototyping purposes without the need to make expensive tooling or relying on expensive craftsmanship. The focus in prototyping stages explains why the technology was initially called Rapid Prototyping. Additive manufacturing has since gained substantial commercial and research traction. AM is growing by 33% a year and is starting to cover all sectors of industry [1].

At the level of research, new processes and material options were developed to tackle the main limitations of the technology: manufacturing speed, mechanical resistance, process repeatability, cost, build volume, and resolution. At the turn of the century, technology became increasingly affordable, open-source, and mainstream. The concept of a machine that can print the parts to replicate itself saw its first examples with Rep Rap and Project Darwin, lowering the entry level for

C. BAÑÓN and F. RASPALL, *3D Printing Architecture*,
SpringerBriefs in Architectural Design and Technology,
https://doi.org/10.1007/978-981-15-8388-9_2

designers and enthusiasts. And, new materials transitioned from research to commercial technologies, expanding the range from polymers to ceramics, metals, biomaterials, and composites.

2.1 3D Printing Technologies: Benefits and Limitations

Generally, 3D Printing presents the following benefits:

- Toolless manufacturing: The parts are directly produced without the need for unique and expensive equipment that requires heavy capital investment and time.
- Bespoke production: The elements can be all different without additional fabrication costs. Consequently, the necessity for standardization is reduced, and individual components can be manufactured to respond to particularized requirements.
- Geometric freedom: Where subtractive methods present challenges with undercuts and complex topologies and casting processes rely on proper draft angles, 3D Printing allows for a substantially larger range of manufacturable geometries.

The main drawbacks of 3D Printing include:

- Printing time: One of the main limitations of additive manufacturing is comparatively slow manufacturing speed. Speed rates vary with the printing technology but are typically several orders of magnitude slower than other industrial processes such as injection molding or extrusion.
- Cost: Additive manufacturing equipment and consumables are typically more expensive than those in long-established technologies.
- Limitations on size: Slow speed and higher costs have a direct impact on the maximum print size. Typical build volumes will be up to $500 \times 500 \times 500$ mm.
- Mechanical performance: While higher performance materials such as metals, technical ceramics, and fiber-reinforced polymers are now printable in high-end machines, the mechanical performance of printed materials is usually poorer than when processed in conventional manufacturing processes.
- Additional substructures: The printing process is done layer by layer, which requires that each one is supported by the previous. In parts that have cantilevering elements or bridges, additional support is needed (except in powder bed processes). This represents extra printing time and post-processing work to remove the support structures.

There are numerous, well-established commercial technologies, with different baseline technology (such as lasers, printer technology, and extrusion technology) and different type of raw material input (such as liquid polymers, discrete particles, molten materials or solid sheets). From the commercially available technologies, the most prevalent in the field of architecture is described.

2.1.1 Fused Filament Fabrication

Invented in 1989 by engineer Scott Crump, fused filament fabrication (FFF) is the least expensive and most widespread additive manufacturing technology. The process uses a wide range of thermoplastics in a filament form. The material is fed to an extruder, which heats and fuses the substance and deposits the melt into the part, layer by layer. The extruder is positioned using computer-controlled actuators according to instructions created by the slicer software. The equipment and post-processing are described in Fig. 2.1.

Substantial disadvantages of this technology are the limited level of detail, determined by the size of the nozzle and print height, and the relatively low mechanical properties of the finished parts, as each layer is joined to the layer immediately below it. However, novel high-performance thermoplastics such as PEI, PAEK, and PPSU are increasingly in demand for industrial-grade manufacturing applications, making them a suitable, cost-effective, and lightweight alternative to some metals such as stainless steel or aluminum.

2.1.2 Direct Metal Laser Sintering

Jointly developed by Rapid Product Innovations and EOS in 1994, direct metal laser sintering (DMLS) is a metal additive manufacturing process that melts and fuses metallic powders in a single process. DMLS forms dense and complex geometries not possible to attain with other metal additive manufacturing methods that use binders or fluxing agents.

During DMLS, a thin metal layer of pre-alloyed powder is spread uniformly on the building plate by two different methods, powder deposition, and powder bed (described in Fig. 2.2). The layer is maintained at an optimum stage in an elevated temperature near the sintering range of the alloy. The high-wattage laser starts moving on a cross-section of the object by heating the powder selectively. A new thin layer of powder is laid out on top of the previous one, and the laser sintering process continues to form the next slice. The fabrication may require additional lattice support to reduce possible distortion and to balance the residual stress. The powder is cooled down once printing is finished (Fig. 2.3). The extra material is then recovered and recycled, leaving behind the final model.

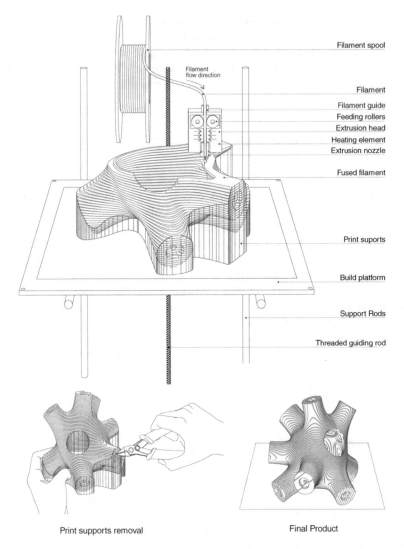

Filament spool

Filament
flow direction

Filament

Filament guide
Feeding rollers
Extrusion head
Heating element
Extrusion nozzle

Fused filament

Print suports

Build platform

Support Rods

Threaded guiding rod

Print supports removal

Final Product

Fig. 2.1 Top: Fused filament fabrication printing process. Bottom: Post-processing of main steps

2.1.3 Binder Jetting Printing

Invented in MIT in 1993, the binder jetting process consists of the selective injection of a binder over a thin powder bed, bonding areas of the powder together to form a solid part, one layer at a time. The powder serves as scaffolding during the printing process (Fig. 2.4). After the de-powdering, the piece has no supports, which constitutes an essential advantage over other printing alternatives (Fig. 2.5). Besides, binder jetting allows for higher printing speed, larger build size, and lower cost.

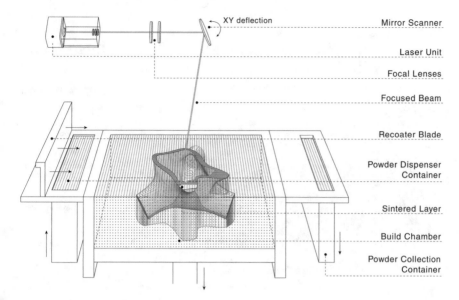

Fig. 2.2 Powder deposition in direct metal laser sintering printing process

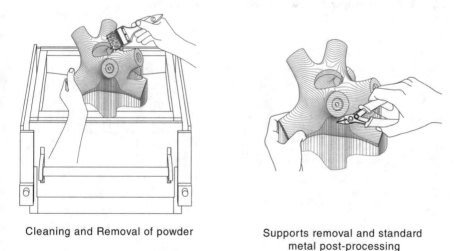

Cleaning and Removal of powder

Supports removal and standard
metal post-processing

Fig. 2.3 Post-processing of DMLS

Parts manufactured using a binder jetting process often present porosity and weak mechanical performance, requiring a post-process to enhance their strength. Materials commonly used are sands, ceramics, and metals in a granular form. The use of sand powder has numerous applications, including the fabrication of fast, inexpensive prototypes, which can be full color, and the production of large molds for metal or concrete casting.

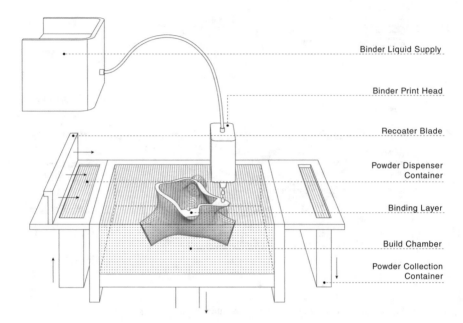

Fig. 2.4 Typical equipment to perform binder jetting printing process

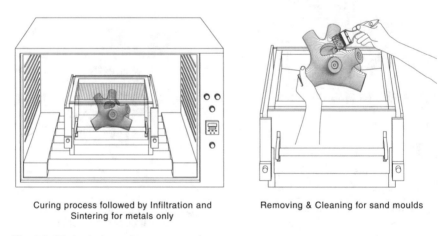

Curing process followed by Infiltration and   Removing & Cleaning for sand moulds
Sintering for metals only

Fig. 2.5 Binder jetting main post-processing steps

Binder jetting of metal powders is an affordable alternative for metal parts, where the printed part undergoes a post-process of sintering and metal infusion to replace the binder for a metal filler.

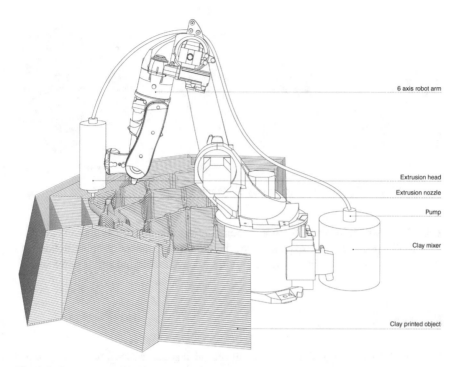

6 axis robot arm

Extrusion head
Extrusion nozzle
Pump
Clay mixer

Clay printed object

Fig. 2.6 Concrete 3D Printing process

2.1.4   *Extrusion-Based Concrete 3D Printing*

Extrusion-based concrete 3D Printing is a large-scale printing technique developed for the construction industry. The printing principle is similar to other extrusion practices, where a large extruder positioned by a robotic arm or a gantry system deposits uncured cement mortar layer by layer (Fig. 2.6). Due to its relatively inexpensive cost, the material can be deposited at large volumes and speed, making it compatible with masonry. Pioneering work on this technology was conducted by Khoshnevis, who named the process contour crafting [2]. In recent years, numerous research groups and commercial startups have advanced the technology, with realized experimental proof-of-concept structures.

2.1.5   *Extrusion-Based Clay 3D Printing*

Extrusion-based clay 3D Printing uses a very similar technology to concrete printing, where a pump extrudes clay into precise positions. The printed part is then air dried and fired in kilns to become a ceramic component (Fig. 2.7). The applications of this technology in architecture were pioneered by Bechthold [3]. The

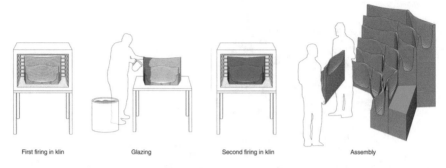

First firing in klin   Glazing   Second firing in klin   Assembly

Fig. 2.7 Main post-processing steps in clay 3D Printing

technology, which instantly became popular among ceramic artists and artisans, has definitively reshaped and expanded the boundaries of masonry construction and opened new paths for experimentation with a considerable industrial potential.

References

1. Campbell I et al. (2018) Wohlers report 2018: 3D printing and additive manufacturing state of the industry: Annual worldwide progress report. Wohlers Associates
2. Khoshnevis B (2004) Automated construction by contour crafting—related robotics and information technologies. Autom constr 13(1):5–19
3. Bechthold M (2013) Design robotics: new strategies for material system research. In: Inside smartgeometry. Wiley, Chichester, pp 254–267
4. Clevenger CM, Khan R (2014) Impact of BIM-enabled design-to-fabrication on building delivery. Pract Periodical Struct Des Constr 19(1):122–128
5. Goldman G, Zarzycki A (2014) Smart buildings/smart (er) designers: BIM and the creative design process. Building information modeling BIM in current and future practices, pp 3–16
6. Kensek K, Noble D (2014) Building information modeling: BIM in current and future practice. Wiley, New York

Chapter 3
Optimized Structures: AirMesh

Abstract This chapter describes the opportunities related to the application of additive manufacturing to the design and manufacturing of non-standard load-bearing structures, focusing on space frames. With 3D Printing, space frames design is not limited by standard components and standard fabrication methods, enabling optimized frames in which the frame topology and nodal joints are specialized for their specific stresses within the structure. Digital design approaches and workflows, including the overall form generation, element definition, and topology optimization are presented in detail. As a case study, the chapter presents an in-depth analysis of the design and manufacturing workflows for *AirMesh*, a stainless steel pavilion developed by AirLab.

Keywords Space frames · Structural optimization · Digital workflow · Complex geometry · Digital architecture

The domain of structural design represents one of the most promising applications for 3D Printing in architecture. The integration of structural analysis and generative modeling is redefining the design of structures. A generative model creates design variations, whose structural properties are computed in almost real time. Numerous variations are automatically created, from which the designer selects the most efficient one. As a result, projects become structurally more efficient yet can be geometrically more complex. 3D Printing comes into play by enabling the manufacturing of intricate and non-standard structural components, where other manufacturing technologies would not be feasible.

As of today, the printing volume and the manufacturing speed constitute a significant limitation for additive manufacturing in architecture. For this reason, structures that require smaller, complex components are more suitable for 3D Printing. Space frames, a structural type consisting of three-dimensional trusses made from linear bars and nodal joints, are an ideal case as complexity is concentrated in the relatively small nodes. They are widely used today in many architectural typologies such as train stations, sports complexes, and museums. A digital design and manufacturing workflow empowered by 3D Printing can redefine the structural possibilities, adaptability, and efficiency of this structural type. With space frames, the combination of inexpensive, off-the-shelf tubular profiles for most of the structure, with comparatively small, complex 3D-printed nodes represents a low hanging fruit for the application of additive manufacturing in architectural structures.

The use of 3D Printing in space frame design provides two key benefits: it allows for freeform structural optimization, and novel aesthetic opportunities. Mass-produced space frame joints available in the market today have a standard, symmetric geometry and are installed throughout a frame disregarding the specific stresses and position in the structure. The joint's form and thickness are not finely optimized, and connection angles are limited. However, with 3D Printing, parts have no additional cost if they are the same or different. So, each node in a space frame can be individually designed for the specific structural requirements, with the number and diameter of converging bars and other technical considerations customized for each point in the structure. Additionally, formal limitations of other manufacturing methods—casting and machining, primarily—are dramatically reduced, allowing for extended design possibilities for the structure's aesthetic expression.

3.1   Structural Optimization of Space Frames

Probably, the earliest example of a space frame (SF) was developed by Alexander Graham Bell in [1]. In his article "The Tetrahedral Principle in Kite Structure" published in the National Geographic Magazine, he described a framework for kites and flying machines "applicable to any kind of structure whatever in which it is desirable to combine the qualities of strength and lightness" [1]. More recently, according to the IASS report in 1984, a space frame is "...a structural system, assembled of linear elements so arranged that the loads are transferred in a three-dimensional manner. Macroscopically, a space frame often takes a form of a flat or a curved surface" [2].

As a system, space frames consist of axial members, which are preferably tubes, and connectors, which join the members together. Currently, space frame nodes are made with conventional manufacturing methods, delivering standardized nodes. The Mero System, illustrated in Fig. 3.1, is a well-established commercial system.

Typically, current SF systems present a regular pattern, primarily due to the constraints of designing with standard elements. As metal 3D Printing eliminates the economic incentives of standard parts, the benefits of a repetitive design are reduced, and the redevelopment of the node details offers new opportunities: custom number and angle of converging maximize frame design, direct printing of threads minimize post-processing machining, and hollow shells of variable thickness can reduce material usage and weight of the node.

Therefore, substantial freedom is added to structural design. There are two scales at which space frames can be optimized: at the frame level and the connector level.

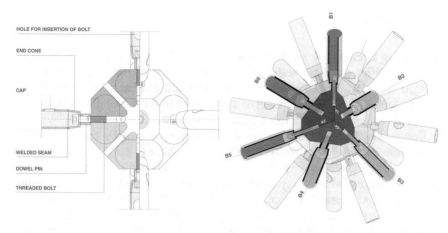

Fig. 3.1 Left: Section of a typical Mero System connector; Right: Sectioned axonometric drawing showing 16 bars connected to a node

At the frame level, the number, position, and connectivity of points can be calibrated to yield lighter and stiffer structures. Combining a generative algorithm, which can propose variations to the frame with a structural analysis module, a ranking of design options can be obtained based on a target parameter such as material consumption, deflection, and cost. At the connector level, the optimization of the geometry can be performed by generating particularized designs that respond to the number, angle, and stresses of the converging axial members, minimizing weight while increasing performance.

3.2   Aesthetic and Technical Opportunities

The contribution of 3D Printing to the design of space frames exceeds structural optimization. Indeed, the expressive and aesthetic possibilities of structures are significant, both at the frame and the node level. At the scale of the whole construction, the structure can adopt any possible form. Therefore, the formal and spatial imagination of designers is not compromised by the structural requirements. Additionally, the customization capacity of printed structures allows for specific technical improvements, such as reducing the need for substructures and interfaces with the other elements that constitute a building.

Due to the unrefined visual quality of existing nodal joints, current space frames are typically concealed or disregarded. However, at the scale of the node, the 3D-printed connectors can be designed to achieve high aesthetic value. Visual continuity between bars and nodes and concealed fasteners, for example, can significantly increase the beauty of structures. Also, custom nodes can integrate more functions such as window frames, enclosure, foundations, floors, and roofs into elegantly designed objects.

3.3   AirMesh

AirMesh is a temporary pavilion located in the Gardens by the Bay Singapore (Fig. 3.2). With a unique design, it demonstrates the potentials of digital technologies for architecture. The research behind the structure has as its main objective the development and demonstration of a seamless and functional digital chain from design to fabrication, with a focus on additive manufacturing.

Conceptually, the pragmatic form of *AirMesh* is defined by the directions of views from the pavilion to the surrounding. The volume is defined by four vectors oriented toward highlights in the landscape: The Dragonfly Bridge, the Silver Garden River, the Marina Bay Sands' rooftop, and the entrance. Four rectangular view frames generate the space, and its faceted form reinterprets a traditional Chinese lantern. The structure lights up in subtle colorful gradients utilizing two-layered breathable porous skin made out nylon netting. The netting transparency allows wind to pass through and the structure to be seen, yet its translucency evidences the pavilion's volume. The view frames serve as openings for the envelope, where the primary views are presented unfiltered. Figures 3.3 and 3.4 illustrate the position of the project in its context, and Fig. 3.5 shows the interior space with the openings framing the selected views.

The pavilion structure is a space frame with 216 bars of different lengths and sections, and 54 unique-printed nodal joints. The design was computationally driven, the manufacturing of parts was digitally produced, and the assembly of the structure was successfully modeled and implemented. The structure weighs 700 kg, and despite its delicate and fine appearance, it can withstand loads 16 times its weight—more than 11 tons.

As a research project, *AirMesh* investigates the structure optimization possibilities related to the design and manufacturing of non-standard load-bearing

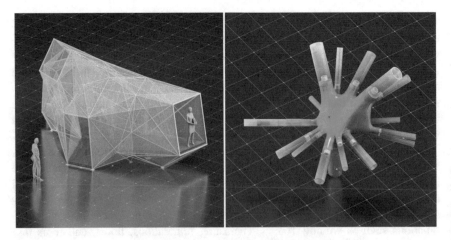

Fig. 3.2 Axonometric drawings of the *AirMesh* pavilion (left) and its joint system design (right)

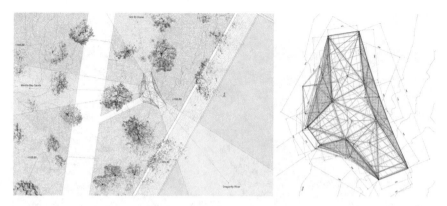

Fig. 3.3 Left: Site plan showing the main four views from the pavilion; Right: Detailed floor plan

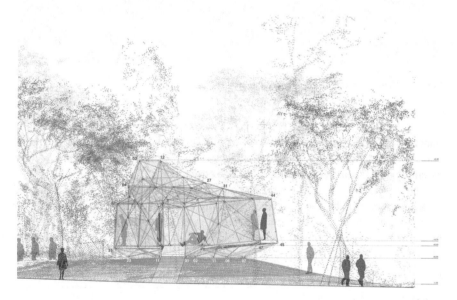

Fig. 3.4 Side elevation

structures using 3D Printing. With this approach, the material allocation is not limited by standard components: Through computational design and structural simulation, the topology of the frame and the form of the nodes are based on the actual distribution of forces in the structure. The integration of multiple functional building components of the architectural system such as foundation, frame, floor decking, façade cladding, as well as lighting features is implemented in the space frame structure. In this way, the spectrum of adaptability of a nodal joint system was tested beyond the structural function and was able to resolve multiple functions in a seamless aesthetic expression.

Fig. 3.5 View from the inside of the pavilion

The high accuracy and precision of additive manufacturing provides the opportunity to develop an innovative joinery system embedded in the 3D-printed components, which resolves the tolerances during construction and improve the assembly process for complex space frame structures.

The structural design follows Eurocode regulations and is the first 3D-printed structure approved by the Singapore Building Construction Authority (BCA) for temporary occupation in Singapore up to three years. The pavilion is also approved by the Urban Redevelopment Authority (URA) for the same period as a validation of metal AM material and SF structure performance over weathering conditions.

Project data

Dimension	$8 \times 5 \times 4.8$ m
Area	28 m^2
Year	2019
Printed parts	54 nodes
Material	Steel–bronze alloy hollow shell, thickness 3–5 mm
Printed volume	8,750 cm^3
Weight	65 kg
Structure parts	200 bars

(continued)

(continued)

Material	Stainless steel circular tube
Dimension	Thickness 1.2–3.0 mm, diameter 20–50 mm, length 0.2–3.5 m
Project team	Lead designers: *Carlos Bañón, Félix Raspall*. Team: *Anna Toh Hui Ping (project manager), Tay Jenn Chong, Wang Sihan, Liu Chi, Huang Kunsheng, Luo Qihuan, Sourabh Maheshwary, Aurelia Chan Hui En*. Consultants: *TCP Engineers, PTL Consultants, WoodFix*

3.3.1 Design

The design of *AirMesh* integrates formal and compositional intentions with structural innovations in a space frame structure using a single, custom algorithm. The objective of the design is to advance the use of additive manufacturing in architecture and through a built project and provide a proof of concept for the 3D-printed space frame system. The design covers a range of structural and functional requirements such as foundation, envelope, and decking. The associated workflow spans from the spatial strategy, the design of the frame's topology and its optimization, to the development of the nodal system.

3.3.1.1 Spatial Strategy

The pavilion is conceived as an inhabitable hollow volume swathed by two thin porous layers. The inner envelope delimits the usable space inside the pavilion, while the outer layer is strategically offset, creating an in-between space of variable depth. The two layers meet in the four view frames defining the shape of the building (Fig. 3.6).

The pavilion served as a testbed for the integration of architectural elements in the SF system. Through relative displacements of the vertices defining the two envelopes, a ramp, a skylight shaft, a sitting area, and a cantilevered window frame are articulated holistically in the design (Fig. 3.7).

Fig. 3.6 Sequence of parallel cross-sections representing the concealed volume between the interior layer (orange) and the exterior envelope (white)

Fig. 3.7 Left: Raised deck view; Middle: Access ramp view; Right: Skylight shaft view

3.3.1.2 Topological Design of *AirMesh*'s Space Frame

The design of the *AirMesh* structure follows a tetrahedral mesh topology. Its bars are configured into a series of adjacent tetrahedrons, which adds significant stiffness not the structure. At the first stage, the project's algorithm accesses functionality from TetGen, a volumetric mesh generator [3]. As input, the code is fed with the target massing for the project, whose volume is subdivided into tetrahedrons. The resulting tetrahedral mesh is used as a preliminary wireframe geometry for the structure (Fig. 3.8).

3.3.1.3 Space Frame Optimization

Once the basic wireframe is set, the overall tetrahedral space frame structure undergoes structural analysis under numerous load cases using Karamba3D, a FEM modeler for Rhino and Grasshopper. The structural analysis data is used to assign optimal bar sizes based on the force distributions within the structure, improving material efficiency, and structural performance. The space frame bars use six types of standard stainless steel circular tubes with outer diameters ranging from 19.1 to 50.8 mm. The wall thickness varies from 1.2 to 3.0 mm (Fig. 3.9).

3.3.1.4 Parametric Design of Nodal Joints

The design of the tetrahedral frame and the optimization of the bar sections provide the key parameters necessary to generate the node's geometry: the number, angle, and section of the converging bars at each node. The algorithm creates the node's

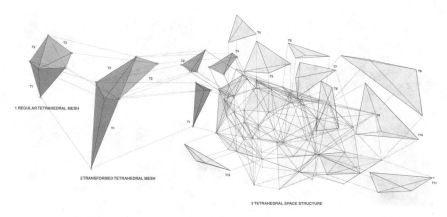

Fig. 3.8   Tetrahedral topology of the generated mesh

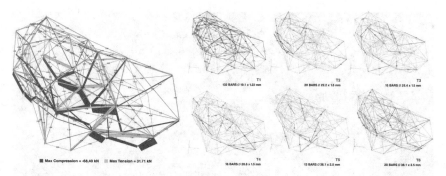

Fig. 3.9   Left: *AirMesh* FEM model and section optimization; Right: Distribution of the six types of bars in the structure

geometry in a series of steps. Figure 3.10 depicts the step-by-step process, and Fig. 3.11 the variations resulting from the different boundary conditions.

First, the length of each branch of the node is calculated. The bar sections and relative angles of all converging bars are evaluated to determine the minimal branch length that avoids collisions. Then, the branch length is adjusted to round up the bar length. The freeform design of the frame entails that each bar has a unique length. To simplify the production of bars, which are done with conventional automation, the lengths of bars are rounded to the closest 50 mm. The variations in length are absorbed by the geometry of the node, which extends each of its branches accordingly.

The next step is the generation of the node's surface. By utilizing the bar sections and calibrated branch lengths, a rough mesh is created connecting all branches with their neighbors. The low-resolution mesh is smoothened using the subdivision surface method. The surface of the node is then thickened by 3–4 mm. Finally, the fastening details are added to the end of the node's branch.

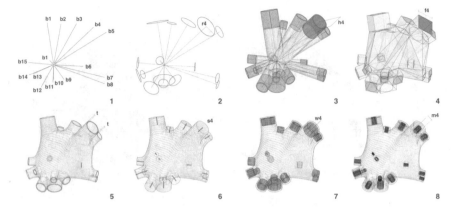

Fig. 3.10 Main steps in the generation process of a typical 3D-printed node

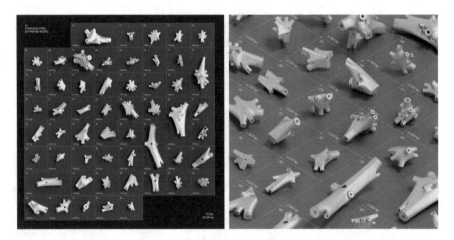

Fig. 3.11 The array of digitally generated nodes

3.3.1.5 Calibration of Nodes and Frames to Reduce Printing Cost

The size of the node is susceptible to the angle of converging bars because when the angle gets very sharp, the node needs to extend to prevent the bars from colliding with each other. Therefore, small adjustments in the frame can represent a significant reduction in printing costs. By analyzing the node design and its tentative cost, the mesh is calibrated to minimize the printing cost. This approach is further detailed in Chap. 8.

3.3.1.6 Nodal Joint Connection System

The fastening detail between bar and node is a crucial aspect of the system. In AirLab's projects preceding *AirMesh*, such as *v-Mesh*, *AirTable*, *Sombra Verde*, and *(ultra) Light Network*, the connection has a peg in the node into which the bar is inserted. This requires careful planning of the assembly sequence to avoid instances in which the bar fits into the two nodes already fixed to the rest of the structure. When this happens, nodes and bars need to be loosened to fit the bar.

The connection detail in *AirMesh* solves this problem. The node hosts the female connections, while the bars contain a bolt inside the bar. Once the bar is in place, the bolt is fastened into the node using an access slot in the bar. The female thread in the node was modeled and printed after a set of experiments to make sure the threads were neither too tight nor too loose given the printer's tolerances. In the case that tapping was needed, a slot in each branch was used to hold the node and align the tapping tool. Figure 3.12 reveals the interior of a typical node.

3.3.1.7 Codes and Regulations

The project was engineered to meet the Eurocode requirements. Therefore, the load hypothesis considering the pavilion loaded to full capacity and high wind loads, safety factors, and calculation methods were compliant with its indications. The results of stresses and deflections complied with the norms and were approved as a functional structure by the Building Construction Authority in Singapore for three years.

FEM simulations of the most stressed nodes were performed, and tensile tests were conducted to validate the specifications of the printed material by the

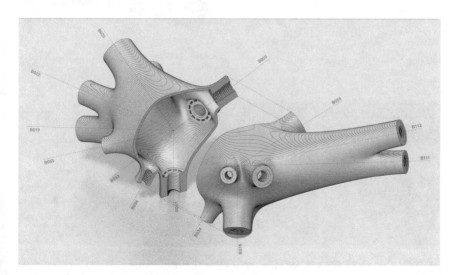

Fig. 3.12 Sectioned axonometric drawing of the node, showing the connection interface

manufacturers. Figure 3.13 describes the forces concurring into a node and the stress distribution. A detailed description of the experiment is shown in Fig. 3.13 as well. The results provided the following properties:

$$\text{Min. yield strength; } py = 455 \text{ N/mm}^2$$
$$\text{Young's Modulus; } Es = 1.475E5 \text{ N/mm}^2$$
$$\text{Density; } Ds = 7,860 \text{ kg/m}^3$$

The bars were made with Stainless Steel Grade 304 with the following properties:

$$\text{Min. yield strength; } py = 175 \text{ N/mm}^2$$
$$\text{Young's Modulus; } Es = 2.0E5 \text{ N/mm}^2$$
$$\text{Density; } Ds = 7,860 \text{ kg/m}^3$$

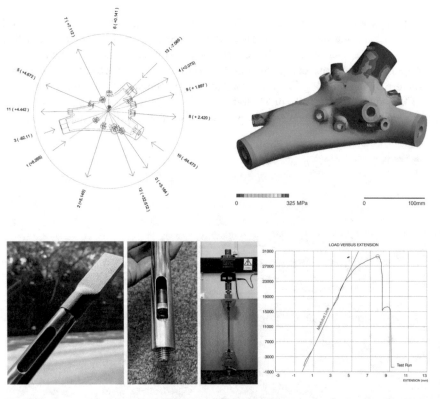

Fig. 3.13 Top: Axil forces and Von Mises stress simulation of a selected node. Bottom: Tensile test of a binder jetting metal 3D-printed node. Peak load = 29.387 kN. Strain at bar = 0.156 mm/mm

3.3.2   Manufacturing

The manufacturing involved four key components: the nodes, the bars, the deck, and the skin.

The manufacturing of the nodes used a metal binder jetting process with curing and sinterization and bronze infiltration post-processing above 1100 °C. After cooled down, the nodes were powder coated in white color and then labeled with removable stickers. The labels indicated the node number, and at each branch, the length and diameter of the bar that is connected to that branch. Each printed thread was manually tested, and those too tight were hand tapped. Figure 3.14 shows a metal node with the threaded connections before powder coating.

For the bars, standard stainless steel tubular profiles were used. Each bar was machined according to its designed length. A slot was machined near each end, designed to access the bolts with a hex tool during the assembly (Fig. 3.15). Then, a stainless steel piece was welded to each end of the holed caps, prior to which two bolts were pre-inserted in the tube. The completed bars were then powder coated in white color (Fig. 3.16).

3.3.3   Assembly

Initially, a simulation of the assembly sequence using a physical model was conducted to understand the behavior of the structure during the erection process. The goal was to work as accurate as possible to create an overview of the structure.

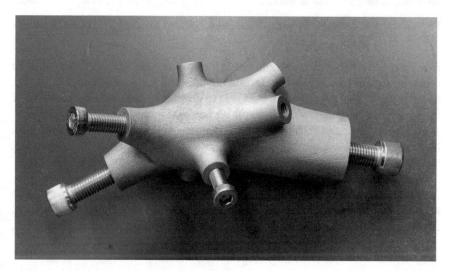

Fig. 3.14 Metal-printed node with the tapped threads

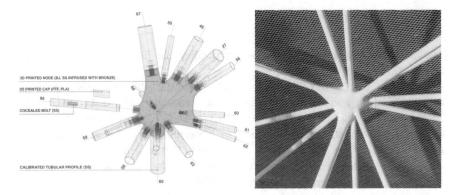

Fig. 3.15 Left: Axonometric drawing of a fully assembled node. Right: Actual joint connected to the structural system

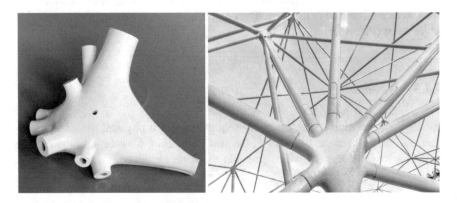

Fig. 3.16 Left: White-coated 3D-printed node. Right: Installed node, including PLA FFF bar caps

However, due to tolerance accumulation, the results were not fully successful in achieving a clear assembly sequence, and digital simulations were required. The 1:1 assembly process was planned in the CAD environment, and, initially, one-third of the structure was prototyped off-site in AirLab's assembly space to test the construction sequence and tolerances. For these, the respective nodes were printed in Nylon PA-12 and assembled with the stainless steel bars. The construction was successful, and the production of the metal nodes was confirmed (Fig. 3.17).

With the metal node, the assembly of the whole structure was tested at SUTD. A custom-made labeling system was developed to identify the components on-site rapidly. The structure was divided into five parts with an identifying color assigned. Equivalent to the nodes, the bars were also referenced with a unique ID number,

Fig. 3.17 1:1 Prototypes assembled in SUTD; Left: Detail of a PA-12 full-scale joint; Center: Close-up of a binder jetting joint. Right: Full view of the structure

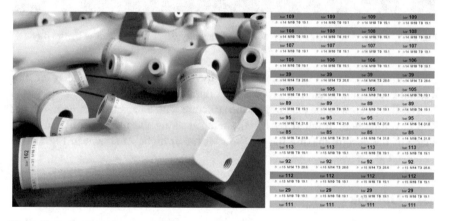

Fig. 3.18 Left: White-coated 3D-printed nodes before assembly; Right: Labeling system

type, size, and nodes to be connected to (Fig. 3.18). The construction took less than two days, carried out by five people using hex keys, ladders, and a laptop computer. The structure was 3D-scanned, and the deviations between the digital model and the scan were compared, giving acceptable results, pondering that the supports were not secured into the ground.

The final step was the disassembly of the structure, its transportation to the site, and the final assembly.

On-site, a concrete slab foundation was executed, leaving open boxes to be cast once the structure was assembled. The final installation of the structure on-site took two days with no issues encountered.

Fig. 3.19 Night view of the pavilion from the Dragonfly River in the Gardens by the Bay

Then, the deck was installed, resting on the bars very near to the nodes, creating a small eccentrical load. The process took one day of work.

The final step was the cladding of the structure with a recycled fishing nylon net. The envelope was installed by hand, stretched along the bars to conform to the shape of the design, making the pavilion visible at night, as shown in Fig. 3.19.

3.4   Outlook

AirMesh demonstrates how additive manufacturing of high-performance materials enable new design processes in which material efficiency is the driver of a new design aesthetic. A set of software packages and advanced 3D steel printing technologies from various technical fields are combined into a single, coherent workflow that enables real-time feedback from engineering and design. This transfer of knowledge and technologies allows for the precise control of the flow of forces and material allocation in space, resulting in a highly resource-efficient structure of unprecedented lightness. The research behind *AirMesh* produced a robust know-how which is applicable to a wide range of building scales and typologies (Fig. 3.20).

Fig. 3.20 Left: Framed view from the inside of the pavilion; Right: Exterior view of *AirMesh*

References

1. Bell AG (1903) The tetrahedral principle in kite structure. National Geographic Society, Washington D.C., pp 14, 219
2. Duggal SK (2010) Limit state design of steel structures, 2e. Tata McGraw-Hill Education
3. Si H (2015) TetGen, a Delaunay-based quality tetrahedral mesh generator. ACM Trans Math Softw (TOMS) 41(2):1–36

Chapter 4
Architectural Ornament: Timescapes

Abstract This chapter analyzes how 3D Printing technologies enable complex geometry projects with an intricate ornamental quality. It speculates about the disciplinary motivations for formal search and the technical aspects such as detail resolution, build volume, speed, cost, and assembly of complex projects. AirLab's built project *Timescapes*, a 113 m^2 pavilion, is analyzed in detail, demonstrating innovations in design, manufacturing, and assembly workflows, as well as the essential challenges and limitations of 3D Printing for freeform ornamental architecture.

Keywords Architectural ornament · Digital design · Complex geometry · Algorithmic patterns · Additive manufacturing

In recent decades, the introduction of digital tools into architectural design has empowered designers with an interest in formal complexity and ornamentation. Algorithmic design methods provide architects with the possibility of creating and mathematically controlling freeform geometries and intricate ornamental motifs with precision, speed, and ease. Furthermore, additive manufacturing delivers the fabrication capacity to bring complex design visions into physical reality. Architects and theoreticians Moussavi [1] and Gage [2] have theorized the integral relationship between digital technologies and the rise of ornament.

3D Printing is central to this expressionist trend. With additive manufacturing, the cost of a part depends on the material and machine usage. Printing a plain or a highly articulated element can have the same cost and take the same amount of time. In contrast, in subtractive manufacturing, the creation of complex geometry and ornament directly translates into higher machine time and, therefore, higher costs.

Digital workflows thread together generative design, digital manufacturing, and assembly processes into a seamless workflow. Architectural ornament is back in the designers' hands, revisiting the ornamental qualities in architecture. And, with additive manufacturing, it is not as a costly endeavor but a legitimate design decision.

© The Author(s), under exclusive license to Springer Nature Singapore Pte Ltd. 2021   39
C. BAÑÓN and F. RASPALL, *3D Printing Architecture*,
SpringerBriefs in Architectural Design and Technology,
https://doi.org/10.1007/978-981-15-8388-9_4

4.1 The Return of the Ornament

At the turn of the twentieth century, the Modern Movement reacted against the excesses of eclecticism and opposed decorative elements in architecture. Inspired by the machine aesthetics of streamlined vehicles, cruises, and planes, and the efficiencies of industrialization and standardization, avant-garde architects proposed an austere vocabulary of simple, unadorned volumes, planes, and lines. Le Corbusier's essays compiled in the 1927 influential book *Toward a New Architecture* [3] provide clear argumentation for this stylistic change that influenced generations of architects.

The modern vocabulary dominated most of the twentieth century until the 1970s when post-modernist ideas questioned the modernist dogma and brought back ornament's communicative power into the architectural discourse. Post-modernism validated the ornamental, while preparing the ground for digital tools' potential for complex and detailed geometries. Computer algorithms facilitated the return of the ornament, becoming the vehicle for the current expressionist trend in architecture.

Two large and often overlapping areas of interest have developed. First, curved geometries have emerged from architects' appropriation of 3D modeling software for animation and vehicle design. Second, an increasingly intricate articulation of projects has emerged from the experimentation with generative algorithms that computationally increase the level of definition of surfaces and meshes to millimetric precision.

4.1.1 Freeform Surfaces

The development of freeform modeling software made the creation and control of complex curvilinear geometries accessible to architects. Nevertheless, curvilinear projects presented new fabrication challenges, as they cannot be fully described with conventional orthographic projections, nor be accurately manufactured with conventional fabrication tools. Accordingly, the digital era entered construction, offering precise machine control directly from 3D files.

Freeform projects also transformed the concept of standardization and tessellation. In straight geometries, a single recurring element, often rectangular or cubic, can be repeated to break down large building sections into smaller, identical components. However, freeform geometry cannot be produced with the same repeating element. Constantly changing curvature, a freeform project requires the creation of multiple bespoke elements.

CNC routing was the first process to address the demands of freeform designs. While being effective in producing multiple unique parts, it is inefficient when routing curved surfaces. A single panel must be carved out from a solid block in a time-consuming and waste-generating process. 3D Printing offers a more efficient and versatile alternative as parts are made layer by layer with little geometric constraints and minimal waste.

4.1.2 Articulated Surfaces

In addition to freeform and curvilinear geometries, computational design enables the creation of an increasingly high granularity of detail. In large surface areas such as those in buildings, the designer cannot reasonably design minute ornamental details manually. Parametric modeling allows the development of algorithms that can automate the creation of detail at a very high resolution on all desired architectural elements at once.

As articulated elements are digitally designed, the manufacturing becomes the main bottleneck. CNC routing allows the creation of details by removing material at the expense of machine time. The finer and deeper the detail, the higher is the manufacturing cost. Additive manufacturing removes this relationship between resolution and cost, as there is no apparent penalty for intricacy. Designers, freed from this fabrication constraint, can explore the ornamental dimension of architecture. Dillenburger and Hansmeyer [4] argued that the confluence of computational design with additive manufacturing is leading to a paradigm shift in architectural ornament.

4.2 Timescapes

Timescapes is the pavilion for the 10th anniversary of Singapore University of Technology and Design (SUTD) (Fig. 4.1). The sculptural structure serves as a time capsule, preserving and displaying the most important artifacts produced by students, faculty, and staff throughout the university's first decade. These objects are stored inside the structure and will be retrieved in 20 years.

The design was conceptualized as a two-sided landscape. Inwardly, it creates an immersive experience by surrounding the visitors with a continuous surface that encloses the interior of the capsule and detaches the space from the vast main hall where it is located. Outwardly, the surface creates new pockets of space for two exhibitions, a performance space, and a welcome bay to receive the visitors.

From a research perspective, *Timescapes* develops design and manufacturing workflows for freeform ornamental architecture. It implements custom parametric modeling tools and low-cost fused filament fabrication (FFF) to create a unique one-to-one architectural space. The project completes the design, manufacturing, and assembly stages, developing the necessary custom tools and workflows. The 113 m^2 pavilion serves as a proof of concept of the practical use of additive manufacturing for cladding layers in architecture.

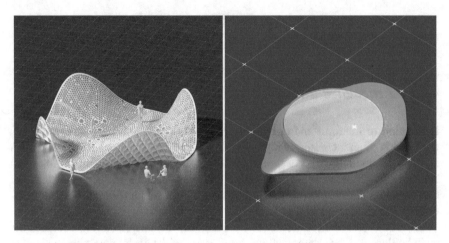

Fig. 4.1 Axonometric drawings. *Timescapes* pavilion (left) and nodal design (right)

The project is made of two fundamental systems: structure and cladding. The support structure consists of an orthogonal wooden grid, whose sections were optimized using FEA analysis and fabricated using high-precision CNC router. Over the plywood structure rests the ornamental cladding, accurately conforming to the designed freeform surface. The surface is made of more than 4,000 uniquely shaped 3D-printed tiles, which precisely fit together and create a continuous and highly articulated surface. The tiles were designed to meet a standard domestic FFF printer build volume using a 3D circle-packing algorithm. Each holds a yellow translucent acrylic, which tints the light and creates a sensorial experience in combination with the wooden structural grid (Fig. 4.2).

AirLab's design and research team worked together with SUTD students and staff to manage the high number of components involved in the fabrication of the iconic pavilion, becoming a collaborative experience for the institution.

Project data

Dimension	12 × 12 × 4 m
Area	113 m^2
Year	2019
Printed parts	4,037 units
Material	PLA
Printed volume	619,000 cm^3
Weight	767.5 kg
Structure parts	47 units of ribs
Material	CNC plywood panel

(continued)

Fig. 4.2 Side view of *Timescapes*

(continued)

Dimension	Thickness 36 mm, Length 3–12 m
Project team	Lead Designers: *Carlos Bañón, Félix Raspall*. Team: *Sourabh Maheshwary, Michele Sodano, Wang Sihan, Liu Chi, Huang Kunsheng, Luo Qihuan, Aurelia Chan Hui En, Jonathon Ng, Anna Toh Hui Ping, Muhammad Syahid Bin Mustapa, Chen Mei Qing Natalie*. Consultants: *WoodFix, Panelogue*

4.2.1 Design

The original brief for the project was to conceal the exhibition underneath the main campus' floor. However, *Timescapes'* design challenged the client's initial expectation by working with a three-dimensional concept. With an undulating geometry, it creates alternative ways for users to relate to the exhibition and introduces a new center in a vast space that otherwise lacks structure and directionality (Fig. 4.3).

The geometry is inspired by the elegance and accuracy of the mathematical sinusoidal equation, which conveys the idea of time, continuity, and infinity. By transferring a 2D equation to a 3D surface, the sine function is transformed into an inhabitable space, engaging users, and conveying timeless values. The surface raises and folds four times, creating four wings and four entrances. Inwardly, it contains a private space while keeping the existing flow of people through the

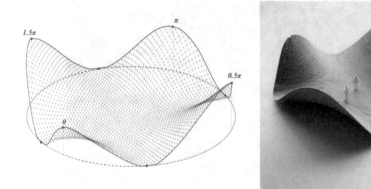

Fig. 4.3 Left: Periodic equation to express of the idea of time. Right: 3D-printed concept model

entrances. Outwardly, the structure adds four pockets of spaces that activate the Campus Center with two exhibitions, a welcome desk, and a stage.

As stated, the installation consists of the plywood grid structure and the 3D-printed cladding. The structure was designed as a waffle system with a square lattice size of 500 × 500 mm. 48 ribs, 24 on each direction, are made of 36 mm plywood and connected through finger joints. The structure's performance was simulated and optimized using an FEA model. A registered engineer endorsed the calculations.

The 3D-printed cladding is made of more than 4,000 unique printed tiles covering 113 m² of surface area (Fig. 4.4). The scale of the project required careful design and planning. In order to meet the tight budget and schedule, FFF technology was identified as both affordable and readily available in printing farms. The design of the tile was adjusted to meet the most standard build volume of 200 × 200 × 200 mm. The approach was to subdivide the double curvature with cells that fit into the printing volume. For this, a circle-packing algorithm was developed, resulting in a tight array of similar yet unique tiles (Fig. 4.5).

The tile design contains several considerations. First, they are designed to be printable without support material, avoiding extra printing, and support removal time. Second, the tiles incorporate a space to fit a circular disk, adding to the expression of the design with a yellow tint and a translucent effect. Third, the connection detail was added to the design: slots conceal a simple yet effective connection method using nylon zip ties, making the process of securing the 12,000 connections fast yet subtle. Finally, a unique ID code was assigned and printed into each tile to identify its position in the whole structure. Figure 4.6 shows the parametric design logic of each tile, and Fig. 4.7, the resulting effect once 3D printed and mounted on-site.

On the flat areas of the structure, where visitors enter and walk through the pavilion, a plywood floor was designed to give continuity to the space. With the same

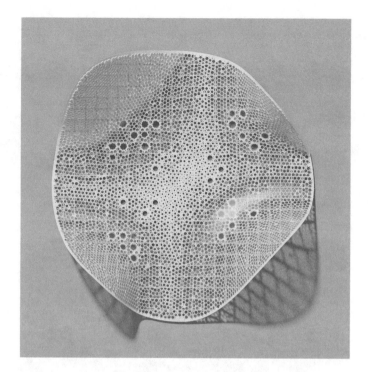

Fig. 4.4 Floor plan depicting the more than 4,000 individual 3D-printed tiles

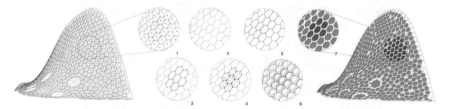

Fig. 4.5 Main steps of the generation of the tiles in one wing

circle-packing method, a polygonal and circular motif blends with that of the printed wings while providing the stiffness necessary for its loadbearing function (Fig. 4.8).

4.2.2   Manufacturing

The manufacturing of *Timescapes* involved three main tasks: the 3D-printed tiles, the flat plywood floor, and the rib structure. The 3D-printed surface constituted the largest and most innovative task in the project. The design of the surface unit tile

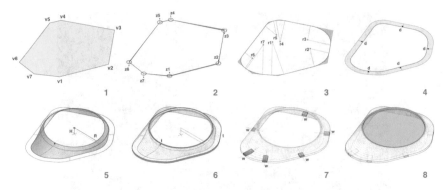

Fig. 4.6 Main steps in the generation of the geometry of the cladding elements

Fig. 4.7 Close-up view of the FFF 3D-printed tiles after installation

required no support structure and fits in a standard entry-level printer. This ensured that each tile could be printed in less than 5 h. The printing was outsourced to a printing company with more than 50 units running in parallel. Within four weeks, at three printing shifts per day, more than 4,000 unique parts were printed without molds or waste. For the same task, a more widespread subtractive manufacturing

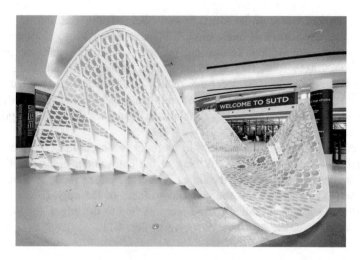

Fig. 4.8 Side view of the surface showing the visual continuity between floor and walls

process would have created a substantial amount of waste, and many of the features of the design would have been lost, such as the embedded connection detail. The tiles were printed using plant-based polylactide acid (PLA), reducing the biodegradation time from hundreds of years to a few decades.

Different AM processes were considered for this project to meet the demanding budget and time constraints. Large-scale robotic printing, stereolithography, and fused filament fabrication were shortlisted. Robotic printing was the most affordable option as it uses plastic pellets, but with single printing equipment at our disposal, the estimated printing time was excessive. Stereolithography was a competitive option with large production capacity and larger build volume, but the increment in cost exceeded the allocated budget. Finally, the manufacturing was made with a printer farm of inexpensive FFF printers.

The pavilion structure was made with 48 ribs, which were machined from 18 mm plywood. Each rib was laminated with two boards with staggering seams, producing continuous ribs with a final width of 36 mm. For the finishing, a whitewash coating was applied.

Finally, to resolve the mechanical requirements and regulations for the walkable surface, a 25 mm plywood floor was chosen. The patterning, resembling the tiled wings, gave continuity to the whole cladding. For the finishing, white epoxy paint was used.

4.2.3   Assembly

The assembly process starts with the waffle structure. 24 ribs in one direction were erected with temporary spacers. Once installed, the perpendicular ribs were fitted into the slots. The waffle was then bolted to the existing floor.

Then, the four wings of the structure were clad with the tiles. For this, the tiles were first attached to each other using cable ties inserted into slots specifically designed for the connection. To facilitate the layout of the tiles, that, for each wing ranged from 700 to 1,400 tiles, a unique ID was printed in the tile, indicating the position of the tile in the structure. Additionally, 1:1 templates were printed, helping unskilled laborers to conduct the task with minimal supervision. Once the tiles were assembled, they behaved like a large chainmail. Figure 4.9 shows the ID printed in the tile and the 1:1 template.

The edge of the wings was made with specially produced tiles. These present an L-shaped section, with one leg attaching to the conventional tiles and the other folding perpendicularly toward the floor. Within the edge pieces, two metal rods were inserted to increase the strength and stiffness of the edge (Fig. 4.10). The assembled edges were installed first. Then, attached to the edges, the wings tiles were "draped" over the structure, as shown in Fig. 4.11.

Fig. 4.9 Unique IDs printed on the tile (left). The tiles are assembled into chainmail (right)

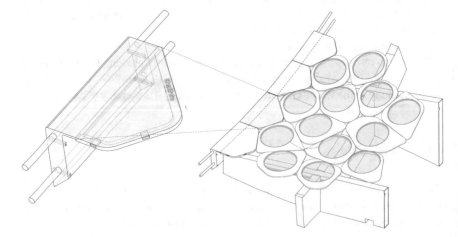

Fig. 4.10 Left: Detail of an edge and connection parts. Right: Integration of components

Fig. 4.11 Process of draping the tiles over one of the four wings of the structure

The final step was the installation of the base. 25 plywood boards were installed with concealed seams within the floor's cellular pattern. The platform components were mounted directly onto the birch plywood ribs with screws. Figure 4.12 shows the base panels before installation.

Fig. 4.12 Image of the assembly process

4.3   Outlook

New requirements in customization, ornamentation, and accuracy are pushing architectural design to a more daring, precise, and demanding level. Additive manufacturing proves to be an emerging technology suitable for highly bespoke and original designs. In *Timescapes*, the use of entry-level printing farms allowed the creation of large and unprecedented spaces (Figs. 4.13 and 4.14). Moreover, as printing technologies show larger volumes, faster times, and lower costs, this experimental use of additive manufacturing may become mainstream. Freeform geometries and ornamental motifs which until recently were difficult to achieve can now be parametrically controlled and manufactured with 3D printers.

Fig. 4.13   Exterior view showing an external exhibition space

Fig. 4.14   Side view showing the interior and exterior qualities of *Timescapes*

References

1. Moussavi F, Kubo M (2006) The function of ornament. Harvard University Graduate School of Design, Cambridge. Actar, Barcelona
2. Gage M (2011) Aesthetic theory: essential texts, 1st edn. W. W. Norton & Co., New York
3. Corbusier L (1927) Towards a new architecture. Payson & Clarke, New York
4. Dillenburger B, Hansmeyer M (2013) The resolution of architecture in the digital age. International conference on computer-aided architectural design futures. Springer, Berlin, pp 347–357

Chapter 5
Architecture Meets Organic Matter: Sombra Verde and White Spaces

Abstract Digital tools deliver high levels of control over a project's geometry and manufactured parts. Digital modeling, CNC control, in tandem with engineered materials such as polymers, plywood, and metal, ensures the fidelity of the physical product against the digital model. However, the dependence of current digital workflows on standardized materials excludes a range of natural materials that can otherwise offer valuable aesthetic and performative opportunities. This chapter examines how additive manufacturing opens new possibilities for unprocessed materials such as bamboo and raw timber. It proposes and reviews a design and manufacturing workflow, in which the specific properties of more irregular materials are digitized so that the specific information of each element is incorporated into the digital design process. 3D Printing becomes crucial, as specialized parts can be manufactured for the unique characteristics of the raw materials. Design workflows, problems, and benefits are presented through two AirLab's built projects: *White Spaces* and *Sombra Verde*.

Keywords Additive manufacturing · 2D scanning · Bamboo architecture · Organic matter · 3D scanning

The use of standardized building materials and components allows designers to elaborate projects without having to consider specific characteristics of each of the components brought to the construction site. Materials are interchangeable; they are expected to be and behave as described in the technical specifications. However, when using natural or lightly processed materials such as raw timber, bamboo, or stones, the architect has no choice but to consider large inconsistencies in the sizes and properties of each element. In current practice, designers working with such materials rely on expert craftsmen who make necessary adjustments to materials to make them fit into the built design as the construction progresses.

Architecture can now embrace design with raw materials and their natural imperfections and bring them into a precise and controlled digital workflow. 2D and 3D scanning, parametric design, and 3D Printing can be coupled, allowing

designers to record the characteristics of each individual material and design and
manufacture components that respond directly to their individual properties.

The research at AirLab investigates the possibilities of additive manufacturing
and associated digital workflows for unprocessed materials through the case of
bamboo architecture. Nowadays, bamboo is becoming a popular material due to its
sustainable properties as a renewable and fast-growing resource. However, its use in
architecture is possible only with labor-intensive construction and expert
craftsmanship. As an alternative, AirLab research proposed a digital workflow for
bamboo architecture. Two approaches were explored, developed, and tested in two
build projects. On the one hand, a parametric design of connectors that embed
tolerances for bamboo poles lengths and diameters was tested in the project *White
Spaces*. On the other hand, the project *Sombra Verde* was experimented with
machine vision to digitize the individual traits of each bamboo pole and design and
print bespoke connectors.

5.1 White Spaces

White Spaces was the Singapore Pavilion exhibited at the Seoul Biennale 2017. The
structure was a lightweight 3D grid of thin bamboo poles. Conceptually, the mesh
filled the 6.5 × 3 × 3 m room into which an inhabitable void was generated
between the access doors (Fig. 5.1). Bamboo was chosen for its associations with
South East Asian architecture, its environmental properties, and its low weight.

The pavilion was designed and manufactured in Singapore and transported to
Seoul in two pieces of check-in luggage. For this reason, the project had to be very

Fig. 5.1 Axonometric drawing. *White Spaces* pavilion (left) and 3D-printed nodal design (right)

light and transportable. Additionally, the design was easy to assemble, as it was built on-site with a group of students from SUTD and Kookmin University.

From a research perspective, the project focuses on the design of a raw bamboo construction system that solves the irregularities of the poles through a slip joint with a clamping detail. In this way, poles of different lengths and diameters could be used within a reasonable spectrum. The use of additive manufacturing enabled the quick prototyping and actual fabrication of 1,000 + functional, complex 3D components in a relatively short period of time.

Project data

Dimension	6.5 × 3 × 3 m
Area	18 m^2
Year	2017
Printed parts	972 nodes
Material	PLA
Printed volume	5,150 cm^3
Weight	6.4 kg
Structure parts	685 bars
Material	Bamboo pole
Dimension	Diameter 8–15 mm, Length 0.3–6.5 m
Project team	Bamboo lead designers: *Félix Raspall, Carlos Bañón, Felix Amtsberg.* Signage design: *Christine Yogiaman, Kenneth Tracy, Michael Budig.* Exhibition curators: *Chong Keng Hua, Calvin* Chua. Team: *Syed Muhd Zabir, Mohd Nazri, Farishiar Bakthiar, Singapore University of Technology and Design Students, Kook Min University Students*

5.1.1 Design

The design process involved the development of a lightweight three-dimensional lattice to fill the allocated 6.50 × 3.00 × 3.00 m exhibition space. Figures 5.2 and 5.3 illustrate the lattice design. The main design challenge was to create an aethereal and highly complex geometrical experience within a tight budget and a short timeframe. The choice of bamboo was a response to this challenge: bamboo has a very good weight-to-strength ratio and is relatively cheap. The benefits are two-fold. First, the structural elements have a very reduced section. Second, the material was locally sourced in Singapore, which allowed prototyping and testing of the structure before it was disassembled and transported to Korea.

Manufacturing a precise lattice structure with an irregular natural material required careful design of the structural system. The key solution was the design of

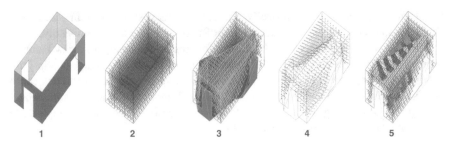

Fig. 5.2 Generation of the geometry of the space. (1) Boundary conditions; (2) Regular grid; (3) Subtracting geometry; (4) 3D Printed nodes; (5) Exhibition boards

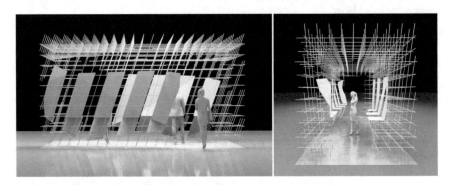

Fig. 5.3 Concept design of the lattice structure

a three-dimensional connector that can handle two uncertainties: the variable bamboo diameter and the variable bamboo length.

The lattice used the thinnest commercially available bamboo pole, with diameters ranging from 8 to 15 mm. For that diameter, poles are around 1 m long, which was a suitable dimension for luggage regulations. The lattice was designed as a skewed cubic grid, where the poles are joined with custom-designed connectors. From the lattice structure, the exhibition panels were hung.

The key component of the project was the design of the printed connector. The geometry resolves the point where poles in x-, y-, and z-directions converge. In contrast with traditional nodal joints, the bars are slightly offset, so that the poles do not intersect but pass right next to each other, close enough to secure them together. This allows the poles to slide before being fixed and enables the use of bamboos with irregular lengths. The poles are secured in place through a thin cable tie, which are threaded through a network of channels that run inside the printed node. Figure 5.4 shows the design of the connector and the actual piece in the structure.

The use of additive manufacturing was crucial in this project. First, the connector design went through several quick iterations that verified geometry and attachment alternatives with precise physical prototypes. Second, the design of the connector with internal channels conceals the fastening system and provides a refined

Fig. 5.4 Node connection detail

aesthetics. The manufacturing of a part with such topology is very impractical for other fabrication methods. Finally, the production scale for 1,000 connectors was small enough to be easily done with 3D Printing but not large enough to justify tooling for other manufacturing processes.

5.1.2   Manufacturing

The manufacturing of *White Spaces* involved two main components: the bamboo poles and the 3D-printed connectors. The bamboo poles were produced in Malaysia. The product comes dried and without branches; therefore, the processing of the poles includes just light washing and painting in white matte pigment. The poles were then bundled into two packages and packaged in plastic film to be shipped within check-in luggage regulations.

The 1,000 + nodes were printed using FFF technology with white PLA material. The final design was optimized for printing, requiring no support. All of the same geometry, the nodes were nested to fill the printing beds of the ten printers that AirLab had at that time. With an average printing time of 15 min per piece, the nodes were fully manufactured within three working days.

5.1.3   Assembly

The assembly of the bamboo lattice was done on-site. The process was sequenced by assembling 24 3 × 3 m planar frames planes on the floor and installing them sequentially from one side of the room to the other, until filling it.

A 1:1 template was printed to guide the assembly of each of the 24 frames. All the frames were indicated in a single drawing to facilitate fabrication. Over the template, the bamboo poles were arranged on the horizontal direction first (x-direction), cutting and extending to meet the design. Then, the vertical poles (z-direction) were assembled and secured using the fasteners embedded in the nodes. Figure 5.5 shows the frame assembly using the template.

Each planar frame was lifted into its final position and attached to an already assembled structure with horizontal bamboo poles that run perpendicular to the frames (y-direction). The position of each node was finely calibrated with a laser level before fully securing the nodes. The assembly process took 1.5 days, with a crew of ten students. Figure 5.6 shows the assembly and calibration process.

Fig. 5.5 Actual bamboo and 3D-printed components superimposed on the 1:1 template on-site

Fig. 5.6 Final calibration of the lattice using line lasers

5.2   Sombra Verde

Sombra Verde develops a new approach to bamboo construction through the application of advanced digital technologies (Fig. 5.7). The core motivation for this project was the search for a construction method that reduces the environmental impact of buildings using sustainable materials while proposing a new language for bamboo architecture.

Sombra Verde addresses the challenges of working with the irregularities of bamboo poles in a different way from *White Spaces*. Rather that creating connectors with embedded tolerance, the project develops a design to a fabrication process that combines computer vision, associative modeling, and additive manufacturing.

Sombra Verde is an urban shelter that incentivizes the use of public spaces in tropical Singapore by protecting them from sun and rain with a fresh and novel expression for bamboo construction. Designed for Singapore's Urban Design Festival 2018 as a resting place in the Duxton Plain Park corridor, the project bridges traditional materials and new technologies, combining raw bamboo poles with individually customized, biodegradable 3D-printed connectors. Despite measuring 8.5 × 6.5 m with a height of 3 m, it weighs less than 150 kg and delicately touches the ground on three inverted tripod columns. Figures 5.8 and 5.9 contextualize the pavilion on its site.

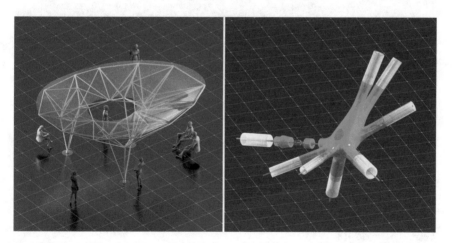

Fig. 5.7 Axonometric drawings of the *Sombra Verde* pavilion (left) and 3D-printed nodal design (right)

Fig. 5.8 *Sombra Verde* bamboo canopy installed in a back alley of shophouses

Project data

Dimension	8 × 6 × 4.3 m
Surface	40 m^2
Year	2018
Printed parts	36 nodes
Material	PLA
Printed volume	23,000 cm^3
Weight	8.5 kg
Structure parts	117 bars

(continued)

(continued)

Material	Bamboo Pole
Dimension	Diameter 30, 40, 50 mm, Length 0.4–3.1 m
Project team	Lead designers: *Félix Raspall, Carlos Bañón, Felix Amtsberg*. Team: *Hu Yuxin, Sourabh Maheshwary, Aurelia Chan Hui En, Tay Jenn Chong, Anna Toh Hui Ping, Wang Sihan*

Fig. 5.9 Aerial view, showing *Sombra Verde* within the green corridor of Duxton Plain Park, Singapore

5.2.1 Design

The design of *Sombra Verde* required the development of a complete construction system and the necessary digital tools. The process included the generation of the geometrical and structural model, the development of machine vision applications within the modeling environment, and the programming of a parametric model for custom nodes and components.

The overall geometry of the pavilion was conceived as a freeform circular canopy. The volume was adapted to provide a soft form that suited the natural environment of the park, starting with a regular radial structural system. Figure 5.10 shows the main generation steps of the tetrahedral mesh. The structure approximates the curvilinear shape with a tetrahedral spaceframe. A parametric model developed by AirLab inputs the structure's wireframe and optimizes the allocation of bamboo poles based on the distribution of stresses within the frame.

The design also included the development of the bamboo structure detailing. This includes the bespoke 3D-printed nodes with concealed fasteners and adaptor

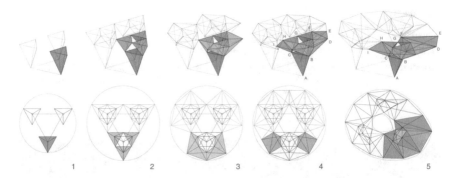

Fig. 5.10 Main steps in the generation of the tetrahedral mesh of *Sombra Verde*

pieces that are tailored to fit each unique bamboo pole. The design workflow for the node, explained in Fig. 5.11, consisted of the scanning of the bamboo ends, the vectorization of the bamboo geometry into a database, and the creation of the connector and node geometry.

A connector piece was custom designed for each bamboo pole, with an automated process that scans each bamboo end section and generates the printing file. The geometry of this piece transitions from the irregular section of the bamboo into a circle. The piece receives a threaded rod and a nut, into which the node will be attached (Fig. 5.12).

The node is designed as a smooth transition between the converging poles. It was parametrically programmed to adapt to any number, angle, and section of converging bars. The node is hollow and contains provisions for embedding galvanized threaded rods, which were inserted at each of the nodes' branch. Specialized nodes were developed to receive the foundations and the roof, with fastening details. Figure 5.13 illustrates the family of parametrically generated connectors.

Special attention was given to the development of the bamboo scanning process. Once the poles were cut to the designed length and labeled, the end sections were

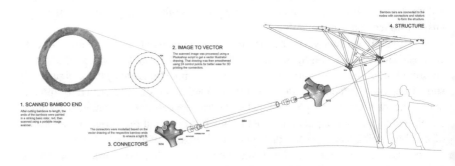

Fig. 5.11 Detail of the connection, showing bamboo pole, end connector, rotator, and node

Fig. 5.12 Left: Scanned bamboo pole section; Center: 3D-printed bamboo cap; Right: Assembled bamboo pole

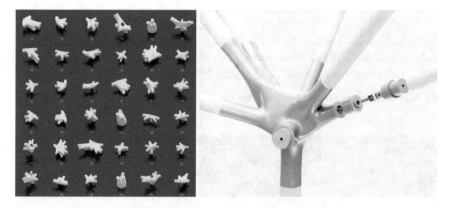

Fig. 5.13 Family of 36 3D-printed nodes

scanned using portable flatbed scanner. The scanned image was then filtered and vectorized with custom routines using functionality from Adobe Photoshop and Adobe Illustrator. Figure 5.14 shows the geometry of each bamboo pole, revealing the variety of sections in a single lot of bamboo. The vector geometry of each pole was refined and used as input to create a custom connector for each bamboo pole.

The mechanical connection between node and bamboo bar was solved with threaded rods and nuts in a concealed detail. The bars contain a threaded rod running through the pole and were secured to the connector pieces. The nodes also conceal a threaded rod going from the interior of the shell and projecting 20 mm outside of the node. The connection between node and bar involved a third piece that using an extension nut and a printed piece that rotates and threads the nut into the rod. Figure 5.15 shows the resulting seamless joinery detail.

An essential part of the design process was the testing of the mechanical performance of the nodes, verifying if the actual tensile and compressive strength matched the assumptions based on specification sheets. The results of the tensile

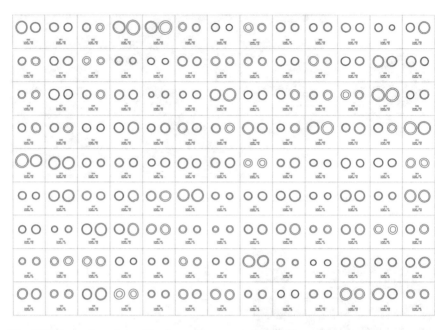

Fig. 5.14 Scanned sections of bamboo poles

Fig. 5.15 Detail of the connection, showing bamboo pole, end connector, rotator, and node

test showed a higher tensile strength expected, even when done in the least favorable printing direction (tensile force applied perpendicular to the printing layers). Despite of this, the decision was made that the nodes should be infilled with epoxy resin, reducing the risks from an invisible printing error. As the structure was placed in a public space for a duration of three months, extra precautions needed to be taken. Figure 5.16 shows the tensile test performed in one of the printed nodes.

5.2.2   Manufacturing

Traditionally, bamboo architecture relies on bundles to spread the uncertainties of each bamboo element into multiple components. *Sombra Verde* proposes a new approach for bamboo construction, replacing bundles with a lightweight bamboo spatial frame. Precise data about each bamboo informs the construction process and enables a new aesthetic of lightness. The manufacturing involved four main components: bamboo poles, nodes and connectors, polycarbonate roof, and foundation elements.

The bamboo poles were sourced from a producer in Malaysia. The supplier provides dried and treated poles, so no additional processing was required. The poles were sorted by diameter and cut to the designed length. Each pole was identified with a unique label for later assembly. Connectors especially designed for each bamboo pole, were printed in PLA and fixed at the corresponding pole using Gorilla glue. As a safety precaution, a threaded rod was installed inside the bamboo, holding both ends together in case the glue failed. The poles were provisioned with a female M6 thread, which was used to fasten the pole into the node.

The nodes were printed using desktop FFF equipment with PLA at 80% infill, as shown in Fig. 5.17. Due to the size and infill of the nodes, the printing time was relatively long, ranging from 70 to 110 h. Once the nodes were printed, a male M6 threaded rod was inserted into each branch and held in place with nuts. Then, epoxy

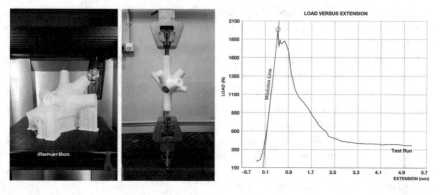

Fig. 5.16 Tensile test for PLA node. Peak load = 1.955 kN. Strain at bar = 0.011 mm/mm

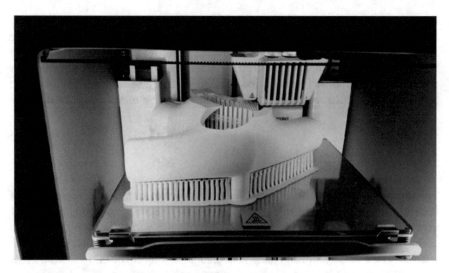

Fig. 5.17 Node printing in desktop FFF printer

resin was cast into each node, bonding together all threaded rods in case that the printing contained invisible defects. Once the nodes were completed, they were labeled with stickers identifying the IDs of nodes and their branches.

5.2.3   Assembly

The structure was assembled like a large three-dimensional puzzle. The sequence was carefully planned in the modeling environment. The structure was divided into three sections, each corresponding to a support, which were assembled separately. A novel connection system concealed the mechanical attachment of the bars to the nodes. Each of the bamboo pole was placed close to the node. Then, by turning the rotator piece, the female and male thread connect and conceal the fasteners. Figure 5.18 shows the set of components involved in the connection of a bamboo pole to a node.

Once the three sections were fully assembled, they were connected into the final annular shape. The complete structure was then propped up, and the tripod legs assembled. First, the structure was fully erected at the SUTD Fab Lab, testing the assembly process in a controlled space before deploying it to the site. Figure 5.19 shows the prototyping at SUTD.

The structure was then disassembled into three sections and moved to the final location. The final installation took less than a day including the installation of the roof that consisted of six polycarbonate panels, and the foundations, made with a steel plate and helicoidal earth anchors.

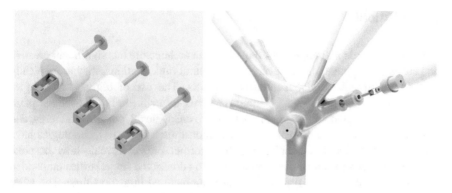

Fig. 5.18 Bespoke components involved in the connection system

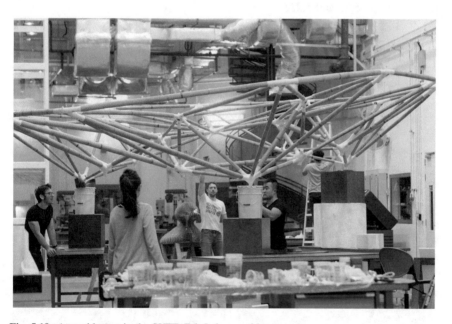

Fig. 5.19 Assembly test in the SUTD Fab Lab assembly space

Sombra Verde was installed at Duxton Plain Park for three months, during which it performed without flaws and showed no signs of wear and tear despite the heavy monsoon season rain and full sun exposure. The pavilion was used as a stage for open-air concerts and created a congregational space as part of the green corridor. The disassembly was as easy as the assembly, conducted in one day and without damage to the pavilion's components, and leaving the site intact.

5.3 Outlook

The use of organic and unprocessed materials in architecture has significant creative and sustainable potential. However, due to natural imperfections, they are typically used in traditional and labor-intensive projects. The use of digital technologies provides new opportunities for natural materials in architecture, using data, parametric modeling, and additive manufacturing to create novel, custom structures, and spaces. *White Spaces* and *Sombra Verde* demonstrate how these technologies have been applied to bamboo architecture. The former project reveals how bespoke complex connectors can be custom designed to address the irregularities embedded in materials and get them prototyped and manufactured in a short time. The latter shows a more advanced approach, where data from the material itself informs the design of components that adapt to the singularities of each of the natural imperfections (Figs. 5.20 and 5.21).

Fig. 5.20 Front view of the *Sombra Verde* pavilion

Fig. 5.21 Upward view from the interior of the pavilion

Chapter 6
Interfacing Printed and Standard: OH Platform, Metadata, and Makerspace

Abstract As 3D Printing finds its way into architecture and construction, how existing materials and building methods interface with new additive manufacturing processes becomes a central area of exploration. Through three built projects, this chapter explores how additive manufacturing can work synchronous with standard materials and existing manufacturing methods. It analyses transitions between printed and non-printed components and describes novel strategies of detail design, hybrid manufacturing processes, and assembly methods that required novel workflows for their materialization.

Keywords Additive manufacturing · Furniture design · Parametric design · Hybrid manufacturing · Exhibition design

3D Printing technologies are already entering the construction industry through experimental projects. How these new construction technologies interface with established processes and how printed components interface with standard ones are important areas of exploration.

Contrary to the conventional design process, where the detailing tends to occur in the construction documents stage, the use of 3D-printed components allows designers to conceive new connection details from an early design stage on. Without the constraints of working with standard parts, the possibilities of designing bespoke details are immense. However, existing manufacturing methods and materials will continue to offer, at least in the foreseeable future, advantages in performance, cost, and volume of production. Hybrid projects combine standard, off-the-shelf elements with specialized printed components, in which the joinery between printed parts and standard materials becomes a critical factor. This requires the precise anticipation of the assembly sequence and the design of how all parts—printed and standard—come together.

A range of design approaches and solutions is described through the analysis of three AirLab projects. In the project *OH Platform*, a freeform surface blends into straight, dismountable aluminum legs. In *Metadata*, flat tabletops blend into

© The Author(s), under exclusive license to Springer Nature Singapore Pte Ltd. 2021
C. BAÑÓN and F. RASPALL, *3D Printing Architecture*,
SpringerBriefs in Architectural Design and Technology,
https://doi.org/10.1007/978-981-15-8388-9_6

bamboo legs. Finally, *Makerspace* proposes an exhibition structure made from structural copper pipes connected with printed connectors.

6.1 OH Platform

The *OH Platform* is a high table designed as a display platform for the 2018 edition of SUTD Open House and the 2018 Venice Biennale of Architecture. The visual concept of the project was to hover a very thin surface over a forest of thin supports. The 12 mm surface was pleated on two corners to increase its rigidity through the form. This simple and elegant design was the response to the very tight budget and schedule given by the client (Fig. 6.1).

The table's top is a thermoformed acrylic sheet that describes a smooth yet tight curvature. This geometric move adds inertia to the table, allowing it to be stiff without a substructure while adding distinctiveness to the design. The surface is supported by 12 thin aluminum legs, clustered in four tripods. This reduces the legs' visual presence. The design of the surface curvature and the positioning of the tripods were defined through a careful structural analysis using FEM. The connection of the linear supports to the surface was resolved with a complex geometry part that interpolates the curvature of the surface of the tabletop and the linear geometry of the legs. The result creates an effect of seamless continuity between all elements. The connecting pieces were 3D printed in high-performance nylon and fitted with a threaded rod that allows the legs to be disassembled for storage and transportation. The tables gently rest on twelve rounded 3D-printed supports adjustable in height that confer great stability to the system while conferring an aesthetic of visual lightness.

Fig. 6.1 Axonometric drawings of two adjacent units of *OH Platform* (left) and a typical 3D-printed node and its connection system (right)

A total of 60 175 × 65 cm tables was produced, allowing different combinations into linear or two-dimensional arrays, where the surface's pleats align and help provide continuity to the extended surface.

Project data

Dimension	1.8 × 0.6 × 0.9 m
Quantity	60 units of exhibition tables
Year	2018
Printed parts	240 nodes 720 units of leveling foot
Material	PLA, PA12
Printed volume	11,300 cm³
Weight	17.5 kg
Structure parts	720 bars
Material	Aluminum circular tube
Dimension	Thickness 4 mm, Diameter 20 mm, Length 0.7 m
Project team	Lead designers: *Carlos Bañón, Félix Raspall.* Team: *Felix Amtsberg, Anna Toh Hui Ping, Hu Yuxin, Tay Jenn Chong, Aurelia Chan Hui En, Sourabh Maheshwary*

6.1.1   Design

The central ideas of the design were continuity and lightness. The tabletop is slender and stiff by introducing a ripple, and the legs are thin relative to their length. The language is one of continuity, where the linear proportion of legs morphs naturally into the surface of the tabletop (Fig. 6.2). With that purpose, the design combines in novel ways and improves existing design and manufacturing methods.

For the tabletop, the design adds curvature using thermoforming. The position and dimensions of the ripple were calibrated to fulfill two requirements. First, the

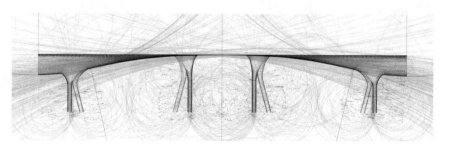

Fig. 6.2 *OH Platform* continuity analysis in section

curvature should add enough stiffness to the surface to resist the loads without substructure. Second, it should not go below the minimum bending radius of the material. Figure 6.3 shows the concept of folding and FEM verifications.

A parametric model for the table was created using a Gauss bell surface. Curvature analysis and structural analysis tools were added to the model to calibrate the main parameters and obtain a design that is both structurally sound and manufacturable using a thermoforming process.

From the rippled surface, the "forest" of thin legs seamlessly emerges. A custom connector piece was designed to create the structural and geometric transition between the surface and three tubular legs. Two types of connectors were needed. The curved type is located right beneath the ripple and the flat type below a leveled area of the table. The shape of both connectors starts tangent to the tabletop surface and smoothly transitions into the three cylinders of the legs. The design embeds a steel threaded rod into the print, strengthening the part and allowing the legs to be dismountable. Due to the geometric and topological complexity of the part and the short production time and volume, additive manufacturing was chosen as the fabrication method.

For the legs, aluminum tube profiles were selected. The end that connects to the table was threaded, while the end that touches the ground was provided with an adjustable printed tip. Figure 6.4 shows the complete assembly of one platform and two back-to-back units.

6.1.2   Manufacturing

The table consists of four main components: the tabletop, the printed connectors, the aluminum legs, and the printed leg tips. All components are shown in an exploded axonometric in Fig. 6.5. For the tabletop, several challenges were present

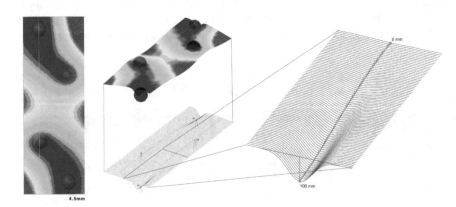

Fig. 6.3 Structural folding principle and 3D model of *OH Platform* tabletop

 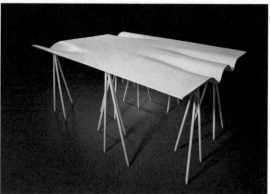

Fig. 6.4 Left: *OH Platform* design; Right: Two adjacent units

at the initial stages of the design. The designed curvature has a small radius of 20 mm, which was beneath the recommendations of manufacturers. Several prototypes were manufactured to demonstrate the bending possibilities of the material with no visible cracks. The production run of 60 tables was done with a single reconfigurable MDF mold that was used for the two types of tabletops with one the specular image of the other.

The connectors mechanically attach the legs and the tabletop. These were printed using FusionJet technology and PA12 material. This powder-based printing process is fast and allows a large printing volume, so, in a single batch, many connectors were printed. Solid-printed connectors being too expensive, the final design proposed shells into which M14 threaded rods were inserted. Then, epoxy resin was cast, bonding the threaded rods and the thin 1 mm 3D-printed shell. Through prototyping, the right geometry, process, and machine setting were established. Figure 6.6 shows some printed shells before the insertion of the rods.

The legs were made of 25 mm aluminum tubes, threaded with female M14 to fit the connectors. Painted white, they blend with the rest of the components. For the tips of the legs, a small adjustable component was printed. The telescopic system allows the calibration of the 12 legs in case the floor is uneven. Figure 6.7 shows a finished node being connected to the tubular supports.

In *OH Platform*, 3D Printing was required because of the complexity of its geometry and the relatively small production volume of 60 tables. However, the short production schedule required the development of a 3D packing strategy to maximize the number of parts produced in a single printing session. As a result, more than 240 nodes and 1,000 leg tips were produced over the course of a week. Production optimization strategies such as these increase the viability of 3D Printing for mass production.

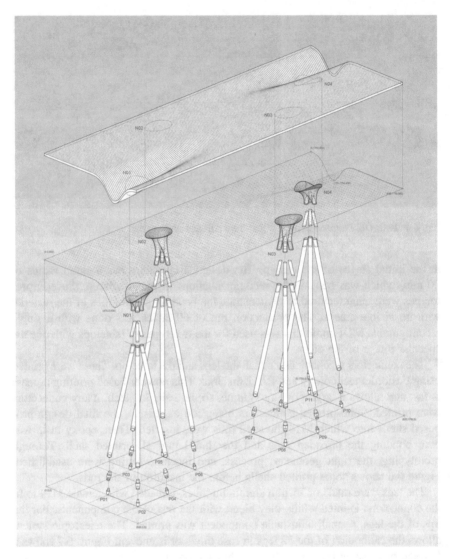

Fig. 6.5 Exploded axonometric drawing showing all of the components of the fabrication of the *OH Platform*

6.1.3   Assembly

The project had a strict budget and schedule, which determined most of the design decisions. Sixty tables needed to be designed for less than 700 SGD each, within three months. Two successful thermoforming prototypes allowed the outsourcing of the tabletops' production. Independently, the design of the connector and legs was

Fig. 6.6 PLA 3D-printed shells manufactured for the connectors

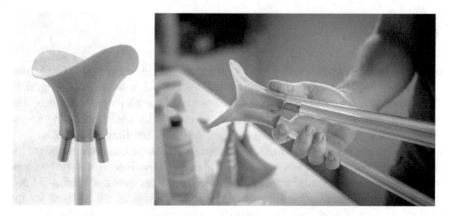

Fig. 6.7 Detail of a completed connector with the embedded threaded rods and the legs

conducted at SUTD. Over ten prototypes were used to refine and validate the aesthetic and functional performance of connectors, which need to attach the curved tabletops to the detachable legs structurally. In addition, the resulting geometry needed to be as smooth and continuous as possible (Fig. 6.8).

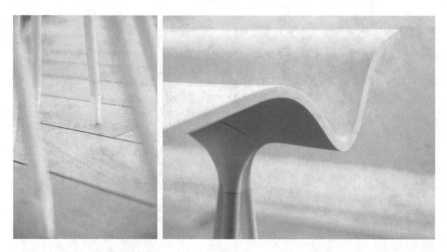

Fig. 6.8 *OH Platform* installed at the Venice Architecture Biennale 2018. Left: Detail of the threaded leg cap; Right: Close-up of the connection between the tabletop and a leg tripod

The assembly of tabletops and connectors was done with structural silicone. The connector to the legs used a threaded connector and the tip was connected to the leg by friction. The assembly of the 60 tabletops, connectors, legs, and tips was done in a single day.

6.2 Metadata

Metadata was an exhibition designed for the 2018 edition of the SUTD Graduation Show. It consisted of 31 tables supported by vertical bamboo legs (Fig. 6.9). The curvilinear form of each table was customized for each project exhibited. The exhibition was dynamic and was rearranged on a daily basis, according to the metadata of each project, including the size, program, and title, among others.

Although the design was simple in concept and form, the customized design, the seamless connection between various materials, and relatively small budget represented a design and manufacturing challenge. The tables were designed, and CNC cut in white laminated plywood and bamboo was chosen as the material for the vertical supports, communicating the sustainability agenda of SUTD. The detailing of bamboo connectors continued AirLab's research on bamboo structures such as *Sombra Verde* (see Chap. 5), which integrates machine vision, parametric modeling, and additive manufacturing. Figure 6.10 presents the exhibition design.

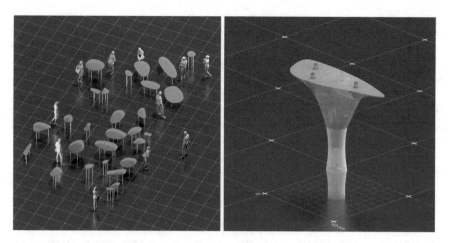

Fig. 6.9 Axonometric drawings of the *Metadata* exhibition (left) and a typical 3D-printed nodal design (right)

Fig. 6.10 *Metadata* exhibition at the National Design Center Singapore

Project data

Dimension	0.5 − 1.5 × 0.5 − 1.5 × 0.4 − 1.0 m
Quantity	31 units of exhibition tables
Year	2017
Printed parts	90 nodes 90 units of foot connectors
Material	PLA
Printed volume	10,450 cm^3
Weight	4.5 kg
Structure parts	90 bars
Material	Bamboo pole
Dimension	Diameter 30 mm, Length 0.4–1 m
Project team	Student Team from the Singapore University of Technology and Design: *Tan Yu Jie, Caroline, Gabriel Chek, Roxanne Then, Chen Jing Wen, Ng Xing Ling, Neo Xin Hui*. Advisors: *Félix Raspall, Carlos Bañón*

6.2.1 Design

Each table has a custom form tailored to the two objects that it was exhibiting: a physical model and a booklet of the project. The form was parametrically created, with a script that took the shape of both objects, generated a convex hull polygon, and created the curve profile of the table automatically. Figure 6.11 shows the design of the 31 tables and the form-generation process.

Based on their size and geometry, the tables had three or four legs, which were made with bamboo poles of different diameters and lengths. A total of 112 bamboo

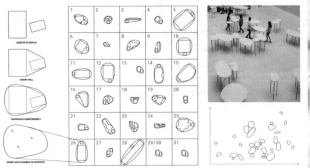

Fig. 6.11 Design of the exhibition

poles were cut to length, and their end sections digitized using a 2D scanner. This information was transferred to a script that assigns each pole to its optimal positions in the table. The algorithm also created the precise geometry of the connector and the print file. The form of the connector blended the pole into the flat surface of the table and included the fastening detailing for easy assembly. The bottom side of the leg included a simple cap, designed individually to fit each unique bamboo pole. Prototyping allowed the calibration of the printing quality, matching the printing density to the shear force requirements of each connector. Figure 6.12 shows the design of connectors, which were customized to each bamboo pole.

6.2.2 Manufacturing

The project consisted of a total of four elements: the tabletops, the connectors, the bamboo legs, and the end caps. The bamboo, sourced locally, was already dried and pretreated. Following *Sombra Verde*'s method (described in Chap. 5), all the legs were cut to length and scanned in 2D. The resulting image was processed and vectorized to generate the particularized models of connectors and end caps. The 112 connectors and 112 endcaps were manufactured in the FFF printers at SUTD's printer farm. To optimize printing speed and costs, the project experimented with different infill levels, increasing the density for connectors with higher requirements such as those with longer and thinner legs. The tabletops were cut with a CNC router, using either a white laminated 18 mm birch plywood for project tables and a 12 mm Corian acrylic solid surface for the welcome desks.

6.2.3 Assembly

The assembly of the table was a simple process. In the case of the plywood tables, the connectors were attached with wood screws using the holes already included in

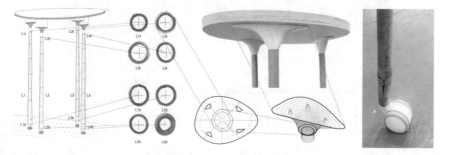

Fig. 6.12 Table legs design using bamboo scanning and digitalization technique

the printed part. For the Corian tabletops, the connectors were glued using cyanoacrylate, which has very high bonding capacity for acrylic-PLA combination. The same glue was used to bond the bamboo to the connector and endcaps. The arrangement of the tables in the space was straightforward. Models and booklets were placed over the custom-designed table (Fig. 6.13). Furthermore, the tables were placed on the location on the site designated by the spacing algorithm that rearranged the projects based on their metadata on a daily basis.

Fig. 6.13 Opening of SUTD Graduation Show. 31 digitally manufactured exhibition platforms display the work of SUTD's graduating class of 2017

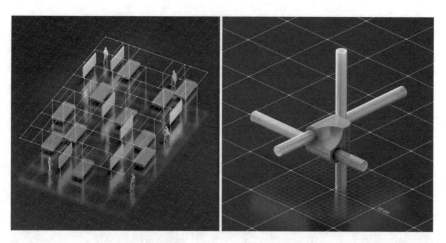

Fig. 6.14 Axonometric drawings of the *Makerspace* exhibition (left) and a typical 3D-printed node (right)

6.3 Makerspace

Makerspace was an exhibition design displayed at the URA Center in Singapore for the graduating architecture class of 2018. The design was based on the concept of reusability (Fig. 6.14). The main elements of the design were done with standard materials without processing. In this way, once the exhibition finished, the materials were ready to be reused. The structure was made with copper tubes connected with custom-designed joints that allow elements to pass close to each other without intersecting. Thus, this connection detail allowed the use of pipes of standard lengths into the designed grid size without cutting pipes to length. Tables and exhibition panels were also made with standard plywood and cardboard stock without cutting or processing. Figure 6.15 shows the *Makerspace* exhibition during the opening night.

This project, together with *White Spaces*—described in Sect. 5.1—focuses on the development of a single slide connector with pre-established angles to produce an inhabitable three-dimensional lattice. Equivalent sections were assigned to horizontal and vertical structural components to achieve a neutral and isotropic condition in the grid.

Project data

Dimension	16.8 × 11 × 2.4 m
Area	185 m²
Year	2018
Printed parts	178 nodes

(continued)

Fig. 6.15 *Makerspace* exhibition during the opening day

(continued)

Material	PLA, PA6
Printed volume	18,950 cm^3
Weight	8 kg
Structure parts	162 bars
Material	Copper circular tube
Dimension	Thickness 1 mm, Diameter 22.2 mm, Length 1–2.4 m
Project team	Student Team from the Singapore University of Technology and Design: *Marianne Ching, Rachel Low, Sng Ei Jia, Marcus Quek, Chan Jia Hui, Natasha Yeo, Pang Yun Jie.* Advisors: *Félix Raspall, Carlos Bañón*

6.3.1 Design

The project proposed an exhibition that minimized waste, which is a notorious problem in exhibition design. Therefore, the main features of the design were created of standard elements, which were not cut nor transformed. In this way, they could be used again or resold upon the end of the show. The spatial organization consisted of an orthogonal grid, where tables and panels showcased the best projects of the year.

The materials included 4′ × 8′ 18 mm plywood sheets, used primarily for horizontal display surfaces, 4′ × 8′ D-board sheets used for vertical display surfaces, and copper pipes, used to create the structure from which boards and lighting were hung. The only customized component was a set of two types of 3D-printed connectors, which connect pipes into the spatial grid. The configuration of the connector is based on a simple filleted cube. The copper pipes almost converge without intersecting, allowing the tubes to continue on both sides of the node without having to cut them to length. Figure 6.16 shows the connector designs.

6.3.2   Manufacturing

Following the design concept, the materials were used in the raw state, so no actual manufacturing was needed beyond the printed connectors. The 178 nodal joints were 3D printed using two printing methods: PLA FFF printing using the SUTD's printing farm and Nylon 6 SLS, using one printer available in the university's Fab Lab. The connector pieces were designed to be printable without supports resulting in a manufacturing time of around one week.

6.3.3   Assembly

The assembly process started with the survey of the grid on the empty exhibition space. Then, the plywood tables were installed, made by the stacking of eight plywood sheets using spacers. The tables served both as a horizontal exhibition area

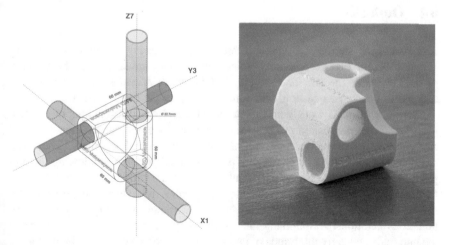

Fig. 6.16 Left: Axonometric drawing of the slip fit connector; Right: PLA FFF 3D-printed joint

Fig. 6.17 Connection of the copper pipes into the node. The black line indicates the target position of the node

and as an anchor for the vertical copper tubes. These were attached to the tables and acted as columns for the suspended horizontal grid. At the tube's top end, the printed connectors were installed. The horizontal tubes were slid into the connectors. The verticality and horizontality of the members were inspected and calibrated using a laser level. Once the grid was aligned, it was fixed in place using cable ties. Figure 6.17 shows the connector and the converging copper tubes during the assembly process. The black line indicates the target position of the node.

The copper structure served as support for vertical exhibition panels and the lighting. The information boards consisted of 18 mm D-Boards and were attached to the copper columns with cable ties. Lastly, the lighting was attached to the horizontal beams, providing the desired lighting for the exhibition.

6.4   Outlook

Architecture is starting to deploy additive manufacturing technologies in construction. As the first experimental projects appear, they are faced with the question of how to merge the new and the old. *OH Platform*, *Metadata*, and *Makerspace* respond to this same question with similar strategies yet three distinct tactics.

The three projects work with the current size and speed limitations of additive manufacturing, proposing projects that mass-produce the larger elements and concentrates the printed components in solving the complex joinery. They take advantage of 3D Printing's capacity to produce quick prototypes, allowing them to produce physical parts fast and empirically test the parts within a short time with the final manufacturing process. Mechanical properties, tolerances, support requirements, bonding capacity, and fit with standard components are quickly refined in numerous iterations. Additionally, the projects explore production runs that are in the hundreds and early thousands of parts, which are the volumes in which printing is still having an edge. Finally, the three designs focus on exhibitions, which can be lightweight and ephemeral and give space for experimentation.

The different nature of the projects requires different tactics. *OH Platform* works with complex geometry, merging a curved geometry into a manifold of legs in a smooth structural connector. To reduce printing time and cost, and increase the mechanical performance, the printed parts hold metal inserts and are infilled with resin, which bond all elements together. *Metadata* employs low-cost printers and, to optimize production, applies different infill ratios based on the structural needs. It takes advantage of the customization capacity of printing, making each table design

Fig. 6.18 *Makerspace* exhibition displaying students' work

and component different. Finally, *Makerspace* explores a three-dimensional structure with adjustable connectors, including the assembly process and calibration into the design of the connector (Fig. 6.18).

Together, the three projects demonstrate how additive manufacturing can work synchronous with existing construction materials and methods, expanding the tectonic possibilities of architecture.

Chapter 7
Systems Integration: (ultra) Light Network

Abstract This chapter investigates the new opportunities for multi-functional parts provided by additive manufacturing in architecture. Building parts layer by layer, 3D Printing eliminates the geometric restrictions of molds or subtractive processes and delivers parts where complex internal cavities and channels are possible. In this way, prospects for integrating multiple technical systems, including structure, power, data, and fluid transmission within a single multi-functional printed part, become evident. This principle is demonstrated with the built project *(ultra) Light Network*, a light-art installation that embeds an electronic light system within a 3D-printed spaceframe.

Keywords Additive manufacturing · Space frame · Light art · Multi-functional part design · Art installation

Applications of 3D Printing in architecture extend beyond optimized structures and ornamental surfaces. New opportunities emerge from integrating multiple technical systems, including structure, power, and data communication in multi-functional printed parts. As the complexity of components can be significantly increased, ducts, cavities, conduits can be embedded into parts to introduce the flow of energy, information, and matter required for the operation of buildings.

Architecture is an assembly of multiple components and systems. In traditional construction, technical systems such as electricity, data, plumbing, and gas are placed in dedicated ducts and concealed behind claddings and in between structural elements. 3D Printing can change the nature of the process as the structure and cladding can be manufactured with all the necessary provisions to house the technical systems. Currently, Building Information Modelling allows designers to integrate mechanical systems into the geometric model and identify conflicts and interferences that will occur on-site. Additive manufacturing goes one step further, allowing the technical systems to determine the geometry of components.

The project *(ultra) Light Network* demonstrates this principle, showcasing how data, power, and light can be embedded in a complex and efficient structure through 3D Printing.

C. BAÑÓN and F. RASPALL, *3D Printing Architecture*,
SpringerBriefs in Architectural Design and Technology,
https://doi.org/10.1007/978-981-15-8388-9_7

7.1 (ultra) Light Network

(ultra) Light Network researches 3D Printing as a means to design multi-functional architectural components that can address not only structural requirements but also power and data transmission and light emission within a seamless and continuous aesthetics. The project produced a light-art installation for the iLight Marina Bay Singapore in 2017 (Fig. 7.1).

Project data

Dimension	10 × 5 × 4.8 m
Area	50 m²
Year	2017
Printed parts	152 nodes
Material	PLA
Printed volume	18,000 cm³
Weight	22.5 kg
Structure parts	715 bars
Material	Polycarbonate square tube
Dimension	Thickness 1 mm, Width 17 mm, Length 0.2–2.3 m
Light source	50,000 units of LED point
Material	Individually addressable RGB LED point
Project team	Lead designers: *Félix Raspall, Carlos Bañón*. Consultants and collaborators: *Mohan Elara, Manuel Garrido, Felix Amtsberg*. Team: *Tay Jenn Chong, Thejus Pathmakumar, Gowdam Sureshkumar, Liu Hongzhe, Mohit Arora, Tan Yu Jie, Joei Wee Shi Xuan, Pan Shi Qian, Goh Yiping, Frankie, students from the Singapore University of Technology and Design*

Fig. 7.1 Axonometric drawings of *(ultra) Light Network* pavilion (left) and a typical 3D-printed node (right)

7.1.1 Design

The project was commissioned through a public competition. The proposal's concept was to create a light-art installation that represents a neural network suspended mid-air, triggering neural activity in response to visitor's actions. The light-weight structure emits light from within, using translucent elements with embedded electronics. The network, inscribed in a prism of 10 (l) × 5 (w) × 5 (h) m, hovers two meters above the public and pulsates with their movement. The lines and connection points contain individually addressable RGB LED points, which are programmed to create dynamic, responsive effects. More than 50,000 individually addressable LED pixels are choreographed by a bespoke algorithm, which reacts to the presence of visitors using sensors located at the bases of the structure. The result is a dynamic, interactive experience. Figure 7.2 shows the installation at The Promontory in the Marina Bay area of Singapore.

The design and manufacturing of the installation were driven by a custom digital workflow programmed within a single parametric model. The design workflow follows a sequence of steps. First, the mesh was generated from the subdivision of the prismatic base volume into smaller tetrahedral units (Fig. 7.3). The resulting wireframe was analyzed and optimized using a finite element method (FEM), making adjustments to the position of nodal joints to reduce stresses and deflections (Fig. 7.4). The structural design challenges the 10 m span by using extremely slender and lightweight elements (30:1 to 135:1 aspect ratio). Unlike other systems

Fig. 7.2 *(ultra) Light Network* installed at the Promontory at Marina Bay, Singapore

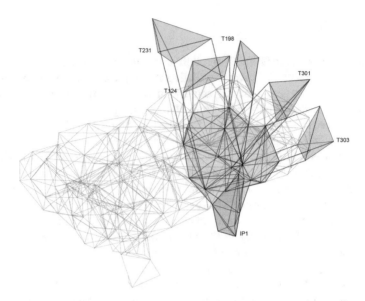

Fig. 7.3 Exploded axonometric drawing of the tetrahedral topology of the space frame mesh

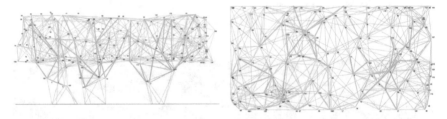

Fig. 7.4 FEM analysis of the force distribution through structural components. Left: Elevation; Right: Floor plan

now in use, its hyper-redundancy enables the structure to absorb stresses and evenly respond to expansion, contraction, and loads such as wind or other eventual punctual forces. For the same reason, the structure is highly resilient as damages to members are absorbed by a mesh that does not contain a single but multiple stable conditions, with an average of ten members converging into each node.

The wireframe and structural information were used to calculate the geometry of the nodes and bars. The nodes were designed considering the number and angle of converging bars. Figure 7.5 shows the foremost step in the generation process. First, the basic, truncated pyramid form was created for each branch. The length was adjusted based on the angle of the converging bars, making branches longer when nodes come closer together to avoid collisions. Then, the solid is transformed into a shell, and analyzing the geometry of the knot, an access cavity was added in

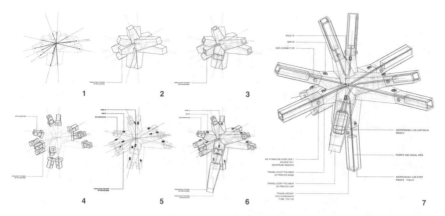

Fig. 7.5 Main steps of the generation of the 3D-printed nodes

the least-crowded quadrant of the node. Fastening details and IDs were finally added to the geometry and the print file created.

The bars were designed as 17 × 17 mm tubes. The lengths were standardized to intervals of 50 mm, to ease production and assembly. To achieve it, the parametric model takes the original length of the bar, which in the structural mesh were of variable lengths, and rounds that length to the closest 50 mm. The length difference is absorbed by the node, which, being 3D printed, does not need to have standard dimensions. The connection detail between the bar and the node was solved with a perpendicular stainless steel bolt.

The suspended mesh touches the ground through three supports. These present a triangular faceted geometry, which blends well with the design of the network. Inside of the support, electronic controllers were installed and concealed behind a polycarbonate cladding. All the elements in the mesh were designed to be translucent polymers, and fit standard LED electronic components. The 715 polycarbonate squared tubes evenly diffuse the light sources, and the 152 unique nodes were 3D printed using translucent ABS and nylon.

The design of the electronic systems was also integrated into the parametric model, allowing to spatialize the position and individual ID of each of the 50,000 light points in space and create coherent animations. This information was exported as a database and included in the light control code. Figure 7.6 shows the circuit design for the installation.

7.1.2   Manufacturing

The manufacturing of the *(ultra) Light Network* structure involves three main components: the printed nodes, the bars, and the supports. The complex geometry of nodes required a hollow interior to house the wiring. Therefore, single material

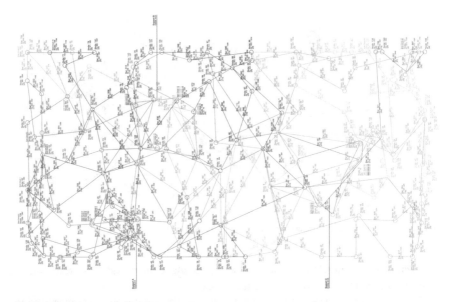

Fig. 7.6 *(ultra) Light Network* circuit design diagram

FFF, the most affordable option, was not possible as the support material would have been impossible to remove. Additionally, the nodes required translucency properties, so printed samples were tested with light sources at a range of luminosities. The final fabrication of the 152 nodes was done with two different technologies: industrial-grade FFF with polycarbonate material for the part and a soluble material for supports using a Stratasys Fortus equipment, and SLS using EOS equipment and nylon six material. Figure 7.7 shows a selection of nodes. The tensile capacity of nodes was tested using destructive methods, demonstrating that the nodes can withstand 200 kg of tensile force before breaking.

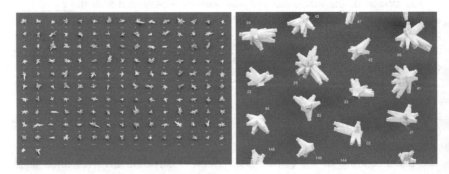

Fig. 7.7 Left: the family of 152 parametric nodes. Right: a detailed axonometric of a hollow node selection

Fig. 7.8 Closeup of the hyper-redundant network with illuminated bars and nodes

The bars are made of standard polycarbonate square tubes with a 17×17 mm outer section and a wall thickness of 1.5 mm. The bars had variable lengths, all rounded to 50 mm. This decision made the cutting process fast, reliable, and precise. A hole was drilled at each end for the mechanical fastener and then LED strips were installed into the tubes. The LED strips were fixed on a rigid translucent rectangular base and covered with a light diffuser that blurs the light points into a homogeneous linear light source (Fig. 7.8).

The base of the structure consists of three simple plywood structures, which receive the weight of the mesh. These were CNC cut from 18 mm marine plywood and cladded with corrugated polycarbonate panels, which recreate a faceted volume.

7.1.3 Assembly

The assembly process required careful planning, as the large volume and 5 m height created instances in which nodes and bars may not be accessible if placed in the wrong sequence. Virtual simulations of the process were developed for numerous alternatives. The adopted strategy was to assemble three large modules off-site that coincided with the location of the three supports in controlled indoor conditions. These, sized to be transportable meeting Singapore road regulations, were brought to site on an open truck and placed in the right position at floor level.

The remaining nodes and bars, with their electronic connections, were installed and completed on-site. Then, the whole structure was lifted to the final height, propped in place while the final supports were installed. Once the floating mesh structure was securely connected to the supports, the scaffolding was permanently removed, and the cladding panels of the columns were installed.

Numerous iterations of the interactive algorithm were subsequently developed. Final calibration of the ultrasonic sensors and adjustments in the color scheme and refresh rate were achieved during the time of the light festival.

7.1.4   Disassembly and Reassembly

After the closing of iLight 2017, *(ultra) Light Network* was disassembled into its nodes and bars and stored. In 2019, it was repurposed as a large-scale chandelier, which was installed at the National Design Center building's atrium for its Open House in 2019. This design, titled *Fractured Light*, was hoisted from the atrium's beams, producing an immersive experience. The hovering mesh tested a different lighting strategy, in which a very bright light source was placed outside the node in a custom-designed, 3D-printed casing. For its fabrication, two versions, one in FFF white PLA, and PA-12 Nylon spray-painted in white were used. The wiring was concealed inside the structure, running inside nodes and bars. Figure 7.9 shows the structure floating suspended at different heights from the ground, and a detail of the light source.

7.2   Outlook

Additive manufacturing opens new opportunities for complex, multi-functional components. The geometric freedom and the level of resolution allow the creation of channels, conduits, and other features that allow mechanical and technical systems to be integrated within single components that address all requirements together. In *(ultra) Light Network*, this principle is demonstrated with printed components that are simultaneously structural nodes, wiring boxes, and light

Fig. 7.9 Above: *Fractured Light*, installed at National Design Center Singapore. Top: The structure is suspended 50 cm above the ground; Bottom left: The structure is hoisted to 12 m height; Bottom right: A close-up of the re-designed lighting source

Fig. 7.10 Closeup of the hyper-redundant network, with illuminated bars and nodes

fixtures (Fig. 7.10). With the ongoing development of conductive printing and shape-changing materials and the advancement of multi-material printing, the opportunities for new multi-functional components has great potential for architectural design.

Chapter 8
Real-time Costing: v-Mesh

Abstract Cost estimation is a crucial dimension of architectural design. Typically, the accuracy in cost estimation progresses from ballpark figures and rules of thumb in early stages to precise quotations as construction starts. Additive manufacturing enables a new approach to project budgeting, as precise costing can appear very early in the design process and provide feedback to the designers when the project is still at a stage where changes are easy and inexpensive to implement. This chapter discusses the new, biunivocal relation between geometry and costing when 3D Printing is incorporated in the design process, and the influence of real-time costing brings to the decision-making process. AirLab's project *v-Mesh* is presented as a case study.

Keywords Cost monitoring · Structural design · Additive manufacturing · Space frame · Realtime costing

In conventional design processes, the construction cost is estimated with a progressive level of precision as the project evolves. Typically, at the schematic design stage, the costing is based on a value per area such as dollars per square foot. In a design development stage, notions of cost per category are presented. It is not until the bidding stage that an accurate understanding of the cost is produced. One of the main selling points of Building Information Modeling technologies (BIM) is that it accelerates the technical definition of the project.

3D Printing further evolve this trend, allowing precise costing and scheduling from early design stages on. Parametric models can generate real-time information about the cost and printing time of a part. This unusual level of precision early in the design process has benefits for designers, as sometimes small changes can have a large impact on cost. The design parameters that have the highest incidence in the project cost can be identified from the digital model, and design adjustments can be implemented immediately. AirLab's project *v-Mesh* demonstrates how the design decision-making process is therefore enhanced, as form and cost are now directly interconnected.

8.1   v-Mesh

v-Mesh (Fig. 8.1) was a pavilion developed for the SUTD Open House 2016. The design was the anchor feature in the event, proposing a 4.2 m high dome supported by three legs, which serve as main exhibition surfaces. Figure 8.2 shows *v-Mesh* in SUTD Campus Center, the university's central atrium space. The structure proposes a lightweight space frame with very slender aluminum bars and bespoke printed nodes. The proposed structural system, which started AirLab's trajectory in this field, was loosely based on Konrad Wachsmann's studies on spatial frames [1]. An integrated parametric model supports the design-to-fabrication approach, in which a robust associative model accurately solves all the physical 3D joints and node-to-bar attachments in real time. The script generates the printing files for nodes and cut sheets for the bars. As costing can be rapidly obtained from the node geometry and print file, the design was adjusted and tuned to reduce the project cost, making the experimental structure fit the project's relatively low budget, reducing the initial cost by a third.

Project data

Dimension	12 × 8.5 × 4.2 m
Area	100 m^2
Year	2016
Printed parts	91 nodes
Material	PA6 (63 pcs), Steel–bronze alloy (28 pcs)
Printed volume	2,950 cm^3
Weight	9.4 kg
Structure parts	369 bars
Material	Aluminum circular tube
Dimension	Thickness 1 mm, Diameter 9.5 mm, Length 0.4–2.3 m
Project team	Lead designers: *Carlos Bañón, Félix Raspall*. Student Team from the Singapore University of Technology and Design: *Mohair Arora, Ryan Chee, Wei Shen, Gammy Chua, Tracy Chow, Ao Chinwen, Pauline Siew, Lisa Koswara, Endy Fitri Bin Saifuldin, Loi Jun, Kai, Tan Jun Kai, Tan Yu Jie, Caroline, Lim Wan Rong, Willa Trixie Ponimin, Rebecca Ong, Liaw Su, Xin, Teo Yu En Dionne, Zhang Ke Er, Shobhaa Narendran, Bryan Lim Wei Guo, Goh Wei Hern, Liew, Sheng Wei Ethan, Tee Yong Kiat*

8.1.1   Design

v-Mesh was conceived as an installation that demonstrates to visitors the state-of-the-art research and technologies at the university and how it envisions new methods of design, fabrication, and assembly. As a prerequisite for this pavilion, the exhibition space was meant to showcase a selection of architectural projects as well as SUTD information materials. Hence, the lightweight structure also served as a

Fig. 8.1 Axonometric drawings of the *v-Mesh* Pavilion (left) and a typical 3D-printed nodal design (right)

Fig. 8.2 View of *v-Mesh* from the main lobby of the SUTD Campus

support for three 5 m long solid flat platforms, which, hovering at different heights, displayed a selection of designs by students and faculty. The slenderness of the pavilion members made the structure almost imperceptible due to its minimal ratio between mass and volume. As such, visitors walked through it when crossing the lobby space and perceived a subtle local change in density, light, and matter to an otherwise neutral and boundless space in the core of the SUTD's Campus.

v-Mesh was AirLab's first project in which an algorithm was able to take a basic form and automatically generate a space frame design, the geometry of bars and nodes, and estimate stresses and cost of the structure. This allowed for real-time design feedback to speed up the design process and meet the client's budget and schedule with a unique project. The design started with the definition of the basic massing: a central dome supported on three exhibition desks. This geometry was subdivided into a tetrahedral mesh, which was used as a wireframe for the structure. The algorithm automatically generated the geometry of the nodes. Figure 8.3 shows the node's geometry and its generative logic.

The cost of each node was calculated based on the volume of printed material and the part's minimum bounding box. Therefore, it was easy to identify where the most expensive prints were located and deduct the parameters that influence the parts' cost. In *v-Mesh*, the nodes in which converging bars reach a very sharp angle must stretch significantly to prevent the bars from intersecting with each other (Figs. 8.4 and 8.5). This increase in the node's volume has a significant impact on its design, and ultimately, this information was used to adjust the structure's design and lower the printing costs.

Once the project budget was balanced, the wireframe was fixed. The final geometry of the nodes was calculated, including the IDs of node and bar, and the cut lists of the bars were updated. Using online printing service providers, the estimation of cost in the parametric model can be easily verified to understand discrepancies and adjust the costing algorithm. These are usually based on printing volume and bounding box and the model's estimation was almost the same as the quotation. Figure 8.6 shows the actual quotation by an online vendor for *v-Mesh*.

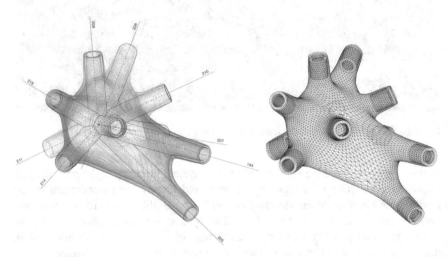

Fig. 8.3 Geometry of each node is parametrically defined according to the angle of each incoming concurring bar. When bars are resulting in sharper angles, the node extends to avoid bar collisions

Fig. 8.4 Cost of each node was dynamically represented to identify expensive nodes and modify the structure's design accordingly. Nodes were scaled by a factor of 4 to facilitate visualization

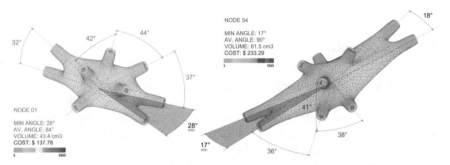

Fig. 8.5 Comparison of cost and volumes between two nodes based on the angles defined by the branches. Angles below 45° are shown

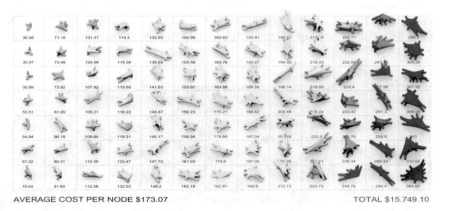

Fig. 8.6 Quotation sheet of the 91 nodes

8.1.2   Manufacturing

The manufacturing of *v-Mesh* comprises two main components: the 91 unique 3D-printed nodes and the 369 bars. For the nodes, several 3D Printing technologies were tested. Initial prototypes were built using desktop FFF printers such as Makerbot. However, the complexity of the node's geometry required significant support structures, and the resulting parts' quality was insufficient for the exhibition quality. For the final production, SLS PA6 polymer and metal binder jetting were chosen, printing 63 nodes in nylon and 28 in a bronze–steel alloy (Fig. 8.7). The printed nodes were then completed with the installation of aluminum pegs at each branch. These were used to insert the bars during the assembly process. The pegs were attached to the node using epoxy glue. Figure 8.8 shows the family of 91 nodes. The bars were cut to length and labeled from 9.6 mm diameter aluminum tubes, with anodized finishing.

8.1.3   Assembly

The assembly process was conducted over five days. During the first three days, a team of three SUTD students assembled the three "legs" of the structure, which support the exhibition tables and hold the "dome" that connects the legs. During the last two days, a team of 15 students assembled the vault-shaped part which bridged between the three legs. The maximum height of the structure was 4.2 m, which required some work at heights with ladders. The connection between the bars and the nodes was resolved by inserting the bar into the peg in the node and bonded with epoxy.

This assembly strategy represented a challenge because the glue has a short open time, and relatively long curing to reach the final cure. That meant that only a few pieces could be placed at a time, leaving them to cure overnight properly. When a more significant section was assembled at a single time, nodes were propped in place with strings and tape to avoid the joints to open. The assembly detail was

Fig. 8.7 Left: the family of SLS PA6 nodes; Right: Metal binder jetting nodes

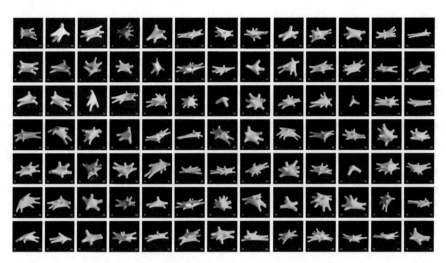

Fig. 8.8 Family of 91 parametrically designed nodes

Fig. 8.9 Structure of *v-Mesh* was extremely lightweight, capturing the ephemeral qualities of the design

refined in later projects, including *(ultra) Light Network*, *AirTable*, and *AirMesh*, where mechanical fasteners were designed. Figure 8.9 shows the final aethereal quality of *v-Mesh*.

8.2   Outlook

The costing process in architecture is one of the most sensitive and make-or-break dimensions of design. Additive manufacturing offers very precise costing for designers, as an accurate budget can be readily obtained from a 3D model. Through computational design, cost information can be successfully used to calibrate and coordinate all other design aspects during early design phases, resulting in more efficient designs. *v-Mesh* demonstrates this trend in action, through a sculptural and structural design that was both groundbreaking and precisely controlled. Through the use of digital workflow and additive manufacturing, the actual cost of the structure was always exact and updated, allowing the design to iterate and converge into a final design that was aesthetically striking, structurally sound, and financially feasible.

Reference

1. Wachsmann K, Burton T (1961) The turning point of building; structure and design. Reinhold Pub., New York

Chapter 9
Assembly Strategies: AirTable

Abstract Additive manufacturing enables the design and manufacturing of complex structures with multiple unique parts. While printed parts can be easily manufactured, the assembly of intricate projects still remains a challenging endeavor. This chapter discusses how 3D Printing gives way to custom connections and careful design of assembly sequences, including the systematic organization of the information, physical parts labeling, joinery strategies, and the creation of virtual assembly models. The project *AirTable* serves as a case study to demonstrate this approach through a built project.

Keywords Additive manufacturing · Parametric design · Furniture design · Space frame design · Metal additive manufacturing

3D Printing enables the creation of complex structures, where standardization of components is no longer necessary. Freeform and ornamental designs as well as projects driven by performance optimization are now supported by a flexible fabrication technology that can translate their singular forms into physical reality. However, as the fabrication of complex components is substantially simplified, challenges in assembly processes multiply when numerous, unique elements have to be put together. The construction of complex structures requires the careful planning of the process, including (1) the systematic organization of the information, (2) the identification of unique physical parts, (3) the careful design of connections, (4) the careful planning of the assembly sequence, and (5) the creation of simplified models to guide the assembly on-site.

Complex architectural projects can include thousands of unique parts. Therefore, to manufacture such projects, the digital model has to name each unique component unequivocally. There are numerous useful naming criteria, such as the part's location in the structure, its type or family, and the assembly sequence, among others. Once the pieces have been appropriately named, the ID has to be included in physical components, which is crucial for organizing the parts during the assembly process. In 3D-printed elements, this can be done in the print itself through debossing and embossing. Useful additional information, such as the ID of adjacent parts, can also be included in the geometry to print.

The design of the joinery is of utmost importance as in architectural projects, the number of joints can be very significant, in the order of hundreds to tens of thousands. Additive manufacturing allows the design of custom connection details, giving designers the ability to create joinery that can provide the desired mechanical and aesthetic results while facilitating the assembly process. Numerous joinery alternatives are compatible with printed parts, such as mechanical fasteners, clipping, and adhesives, among others.

The assembly of a complex architectural project presents challenges that, if not carefully considered, can delay or hinder the process. Accessibility, transportability, scaffolding, and weight have to be considered. Simulations of assembly scenarios can be adequately planned and comparted in the digital model, making sure that each part or sub-assembly can be transported to its final location, placed in the right position, and fastened to the rest of the structure. A final consideration is the creation of proper documents to correctly and efficiently assemble the components. Ad hoc 3D models, where the parts' IDs and assembly sequences are easily accessible, are crucial. 1:1 templates, scale models, and 2D drawings can complement the construction documentation.

An in-depth discussion of the project *AirTable* serves as a demonstration of the challenges in the assembly of a complex three-dimensional structure and the strategies to overcome them.

9.1   AirTable

AirTable (Fig. 9.1) is a large meeting table commissioned by the Center for Digital Manufacturing and Design (DManD). Located in its main entrance, the table represents the innovative and experimental spirit of the center, which specializes in advanced digital manufacturing processes. The project creates a very slender and intricate structure in a design that can only be achieved thought digital design and additive manufacturing. The diameter of the structural elements was pushed to the minimum. Using 6 mm tubes, it required a bespoke connection detail that conceals the fastening features. Figure 9.2 shows the radial design of the *AirTable*, and Fig. 9.3, the structure in the DManD Center.

The project's connection details and the design of its assembly sequence were subject to additional research as the millimetric precision was put to test in a hyper-redundant structure.

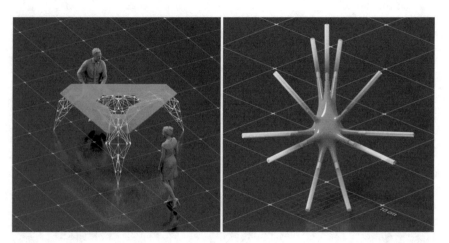

Fig. 9.1 Axonometric drawing. *AirTable* design (left) and a typical 3D-printed node design (right)

Project data

Dimension	$3 \times 3 \times 1$ m
Quantity	1 unit of high table
Year	2018
Printed parts	84 nodes
Material	Steel–bronze alloy hollow shell, Thickness 3 mm
Printed volume	1,000 cm^3
Weight	7.4 kg
Structure parts	462 bars
Material	Stainless steel circular tube
Dimension	Thickness 2 mm, Diameter 6 mm, Length 50–550 mm
Project team	Lead designers: *Félix Raspall, Carlos Bañón*. Team: *Tay Jenn Chong, Anna Toh Hui Ping, Aurelia Chan Hui En*; Students from the Singapore Polytechnic: *Ye Jia Jie, Jona Lim*; Students from the Temasek Polytechnic: *Syed Muhd Zabir, Mohd Nazri, Nurin Farishiar Bakthiar*

9.1.1   Design

Designed for a versatile workspace, *AirTable* can accommodate multiple use requirements such as hot-desking and meetings for nine people. The relatively tall height of 1 m allows for a standing desk position or sitting position with high stools. Its triangular shape ensures that six users face each other comfortably. Visually, the large table works with the contrast of the white and solid tabletop supported by a thinnest network of black lines. The whole weight of the desk is distributed through the mesh into three minute points of contact with the floor.

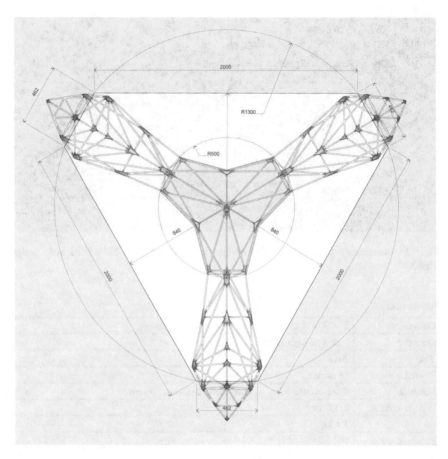

Fig. 9.2 Floor plan of *AirTable*

Figure 9.4 illustrates the design of the structure and the leg detail. Besides its daring look, the reduced footprint for the table legs leaves ample legroom to increase the users' comfort. In addition, the positioning of the legs beyond the tabletop's outline maximizes the usable tabletop space.

Through a generative wireframe model, the placement of the structural nodes was done according to the structural performance using FEM. In an iterative optimization process, the final design is a sturdy tetrahedral mesh with 84 nodes and 306 bars.

The design of the nodes defines a smooth geometric transition between the converging circular sbars using a curvilinear form. The structural connection between nodes and bars was particularly challenging since the 6 mm diameter of the bars left little design space. Several versions were tested, including adhesives, pins, and threads. The final design involves the use of left- and right-hand threading, so that each bar, when rotated, screws simultaneously into the nodes at

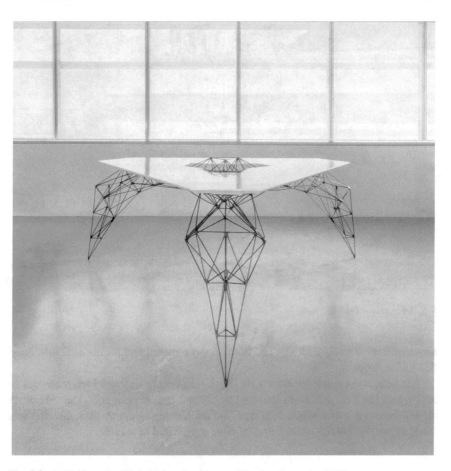

Fig. 9.3 *AirTable* at the Digital Manufacturing and Design Center in SUTD

both of its ends. Nodes use a male M4.5 thread, while the bars use a corresponding female thread. Figure 9.5 shows a detail of the node and an early prototype with a perpendicular peg connection.

9.1.2   Manufacturing

AirTable consists of three main components: the printed nodes, the bars, and the tabletop. The nodes were manufactured with metal binder jetting technology, with post-processing sinterization and bronze infiltration. A hole was designed to allow for powder removal after printing. The threads in the nodes were hand tapped using left- and right-hand dies. The direction of the thread in each branch was physically identified with a small protrusion to indicate that the thread was left-hand. Early

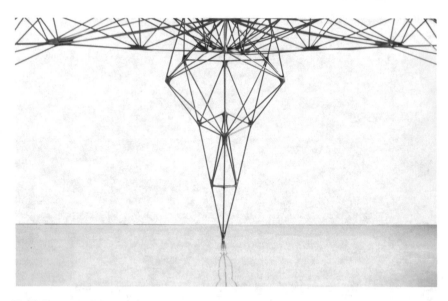

Fig. 9.4 Detail of the support

prototypes were done with DMLS, but support removal and higher cost dismissed this technology.

The bars were made of 6 mm round stainless steel tubes, which were cut to length and hand tapped using left- and right-hand taps at each end. The side using the left-hand tap was marked with tape, to facilitate the assembly. The tabletop was made of acrylic solid surface material over a plywood structure.

9.1.3 Assembly

The assembly sequence required careful planning, as the direction of assembly determined the ease or difficulty of the connections. The chosen order began from the center of the table and radiated outwards, one bar at a time, toward each leg of the table. The assembly, tuning, and finishing of the structure required approximately 30 days to complete by one full-time researcher with intermittent help from interns. The inherent stiffness of the growing connected geometries made it easier to propagate the connections in an outward direction than to join three completed sub-assemblies together. To achieve the perfect fit that the whole system was designed for, each bar had to be carefully calibrated as they needed to rotate into two nodes simultaneously. The erection of the structure was a precise operation, and patience and attention to detail were critical. Figure 9.6 shows the selected

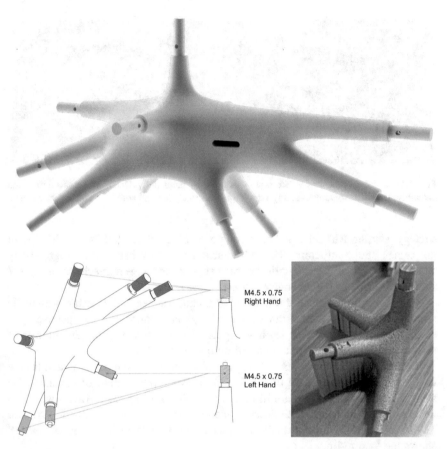

M4.5 x 0.75
Right Hand

M4.5 x 0.75
Left Hand

Fig. 9.5 Top: 3D model of a node in the *AirTable*. Bottom left: Detail of the final fastening system; Bottom right: and an early prototype using DMLS with a perpendicular peg connection

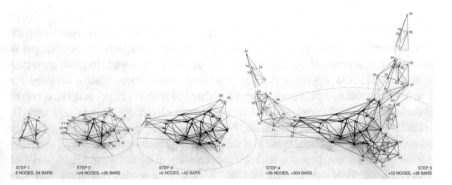

| STEP 1 | STEP 2 | STEP 3 | STEP 4 | STEP 5 |
| 6 NODES, 24 BARS | +24 NODES, +56 BARS | +6 NODES, +42 BARS | +36 NODES, +304 BARS | +12 NODES, +36 BARS |

Fig. 9.6 Assembly strategy showing five main steps, starting from the center of the structure toward the perimeter

Fig. 9.7 Assembly process for one node was done by hand (left). Detail of a node during the assembly, in which some converging bars have been tightened and some are still to be fastened

strategy, starting with an initial core composed of six nodes and 24 bars located in the center of the structure. The subsequent nodes and bars were progressively attached in a radial fashion until the three support nodes were reached. Figure 9.7 shows two steps of the assembly process.

With the aim to reduce the number of unique elements, the design was made radially symmetrical, where three identical legs are rotated 120°, respectively, to each other. In addition, each leg has a specular symmetry, reducing the number of unique nodes to 28 and the number of unique bars to 62.

The table spans a total of 3 m, weighs 108 kg including the tabletop, and is able to withstand a uniform load of 2 kN with an average deflection of 6.4 mm. The slender stainless steel structure has a total weight of 14 kg. A load test was conducted on the table with three different load cases: 120 kg, evenly distributed, and 120 kg of punctual load near the leg and in the middle of the span. Figure 9.8 shows the load testing.

9.2   Outlook

As architecture enters the era of 3D Printing, the manufacturing of complex parts becomes simpler. However, with increasingly complex projects, the challenges of assembly become apparent. Coincidentally, digital design, and additive manufacturing provide new opportunities for enhanced construction. Custom connection details that facilitate assemblies, such as embedded threads, clips, and slots, can be embedded in printed parts. Additionally, parts can physically contain tags and IDs that identify the location, installation order, and orientation that can make the assembly of complex projects easy to follow. Simulations of the assembly sequence can secure a smooth process without accessibility, tolerances, and transportation difficulties. *AirTable* demonstrates how a careful assembly planning can take advantage of additive manufacturing possibilities and deliver an apparently seamless structure with elegant concealed structural connections (Figs. 9.9 and 9.10).

Fig. 9.8 Load testing was conducted using a 120 kg distributed weight

Fig. 9.9 *AirTable* at the Digital Manufacturing and Design Center in SUTD

Fig. 9.10 Close-up views of
different 3D-printed nodes in
AirTable

Chapter 10
Outlook

Abstract This chapter discusses ongoing and future directions of 3D Printing in architecture through a reflection on the topics presented in this book. It summarizes the key existing and emerging applications. It describes the new design workflows including structural modeling and optimization, custom connection design, system integration strategies, manufacturability assessment, and assembly planning. Finally, it concludes by discussing three upcoming topics in 3D Printing: artificial intelligence for 3D part design, large-scale printing, and computer-assisted assembly processes.

Keywords Additive manufacturing · Computer-aided design · Design to fabrication · Design workflows

Additive manufacturing is gaining traction across industries as an emergent and disruptive technology. The era of 3D Printing for prototyping is long surpassed and is already the preferred fabrication choice in specialized applications where complex geometry and small batch production are strategic. For example, the aerospace sector requires complex ultra-high-performance parts in small production batches, while medical applications require personalized pieces tailored to each patient. However, as printing becomes faster, cheaper, more reliable, and larger, more industries start to implement and benefit from additive manufacturing.

Architecture is no exception. Despite its large size, buildings are typically singular objects, so 3D Printing's customization capacity provides an advantage. Additionally, preferences for intricate geometries and highly detailed architectures demand unique, high-complexity parts, where 3D Printing offers many opportunities. The application of additive manufacturing in construction is still driven by experimental projects that explore how this technology empowers designers by enabling the fabrication with forms and materials that were unimaginable before.

Empowered by digital design and additive manufacturing, two distinctive architectural trends recognize that formal complexity can have both performative and aesthetic advantages: the performative and expressionist trends. Specific applications become evident, including furniture, interiors, and pavilions. And, new

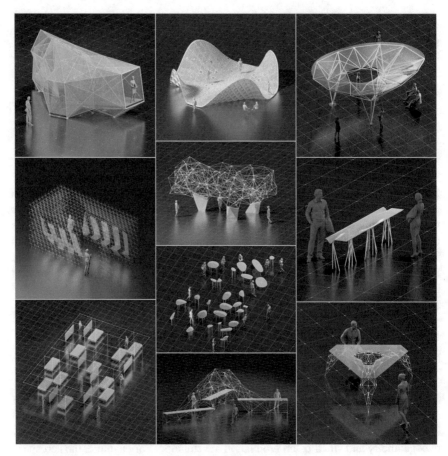

Fig. 10.1 Ten AirLab projects demonstrate the use of 3D Printing in architecture in a range of applications, from furniture to structure, and in response to performative and expressive trends

design workflows are proposed, developed, and tested. Through ten projects, AirLab reviews these trends, applications, and workflows (Fig. 10.1).

10.1 Key Applications

Despite the clear benefits of additive manufacturing in supporting geometric freedom and one-off productions, it still presents significant disadvantages in terms of size, time, cost, and performance. The design research work of AirLab presented in the book identified two critical areas of application that are compatible with the additive manufacturing's constraints and developed custom workflows, working prototypes, and demonstrative projects.

10.1.1 High-Performance Structures

The domain of high-performance structures becomes a unique testbed by combining performative requirements and expressive possibilities. The current computational capacity means that the stresses inside of a structure can be quickly computed and options compared in the search for optimal distribution of materials. In return, the structures acquire a high level of intricacy and lightness that offer new aesthetic potentials. The projects of AirLab presented in this book have refined high-performance structural design through the seven projects presented in the book: *v-Mesh*, *(ultra) Light Network*, *White Spaces*, *Sombra Verde*, *AirTable*, *Makerspace*, and *AirMesh*. In these, the development of a seamless design and manufacturing workflow and the adoption of 3D Printing technologies yield structures that are surprisingly efficient, lightweight, and elegant and would be unachievable otherwise.

10.1.2 Freeform Articulated Surfaces

The formal and material possibilities of 3D Printing are capitalized beyond structural efficiency, engaging the expressive dimensions of architecture. Geometry and ornament become topics of exploration in their own right. Three projects by AirLab focused directly on the expressionist trend: *OH Platform*, *Metadata*, and *Timescapes*. These demonstrate that freeform surface modeling, combined with additive manufacturing, can derive new aesthetic dimensions for architecture and design.

10.2 Enhanced Workflows

Digitally driven workflows empower designers to work on different technical aspects simultaneously and within a single modeling platform. Computer simulations provide information on performative and manufacturing variables in real time. This is used to iterate the design, either manually or in an automated manner. Additive manufacturing is a powerful addition to the digital workflow, as manufacturing information such as printability, cost, time, and support requirements can be accurately predicted and fed back into the design. Additionally, the manufacturing process can be directly integrated, obtaining the printing files automatically. Five key functionalities of the digital workflow have been identified and developed. Figure 10.2 shows a diagram of the design workflow for the project *AirMesh*.

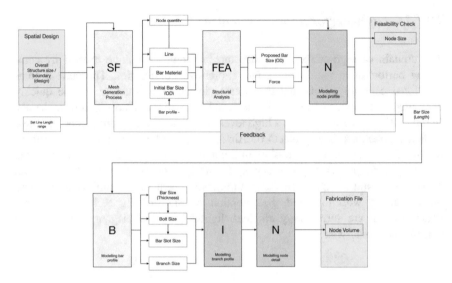

Fig. 10.2 Diagram of the digital workflow in *AirMesh*

10.2.1   *Structural Modeling and Optimization*

The digital workflow can produce structural information directly within parametric models. The geometric information of buildings can be augmented with loads, materials, and sections, obtaining the force distributions within the parts. This information serves to validate the design's structural integrity and refine it accordingly. Often, this process is done in an automated manner. At AirLab, geometric modeling is primarily done with Rhinoceros and Grasshopper; therefore, FEM software compatible with these modeling platforms such as Karamba3D is preferred. This favors the creation of an interconnected workflow between design modules where the flow of forces and structural data are used at many stages of the design process.

10.2.2   *Custom Connection Design*

Digital design and additive manufacturing make the creation of complex parts and assemblies an important area of innovation. Taking advantage of the new freeform possibilities, the connection details can be custom designed on each project. Threads, clips, and slots, among others, can be directly added in parametric models and experimented within building components.

Projects at AirLab rethink and refine all connection details from early stages in the design process on and refine them in multiple physical prototypes using additive manufacturing. This detail-oriented approach would be inconceivable with traditional manufacturing methods, where bespoke designs need to be communicated to and prototyped by craftsmen and take a long time.

10.2.3   Systems Integration

With additive manufacturing, technical systems such as wiring for power transmission and data communication can be included in the project's geometric model at the early stages. This information can be used to create components that include provisions for lighting, wiring, and fluid transmission, for example. 3D Printing parts offer no penalty for complexity as long as it is printable, making the integration of functions into single multi-functional elements a thought-provoking option, and an important area of exploration.

10.2.4   Manufacturability Assessment

With parametric modeling, information concerning the manufacturing of parts can be integrated into the digital workflow. Accurate estimations of the printing time, material usage, parts' cost, requirements for support, and printability can be obtained and used as feedback to refine the design from early conceptual stages on. In this way, the design can be optimized to reduce cost and time and increase manufacturability. Additionally, the parametric model produces the print file directly, making the production of hundreds to thousands of unique parts a relatively simple endeavor.

10.2.5   Assembly Sequence Planning

Typically, complex projects that fully utilize the potentials of additive manufacturing consist of multiple unique components. The fifth functionality embedded in the digital workflow is the planning of the assembly sequence. The step-by-step process must be studied in the digital world, to verify if the installation of each component at each moment is feasible and safe. The results of the simulations can be recorded in animations and drawings used on-site during the assembly. Machine learning algorithms in combination with sophisticated 3D models can be crucial to solving complex assembly problems. Figure 10.3 shows ongoing experiments by AirLab combining a deep reinforcement learning engine with a robotic arm to develop logical structural assembly sequences.

Fig. 10.3 Left: Virtual physical engine for machine learning training; Center: Robotic arm with bespoke pick-and-place end-effector; Right: Final assembly

10.3   Becoming Mainstream

Progressively, experimental projects demonstrate that additive manufacturing is a promising technology for architecture. However, to become a mainstream construction method, existing building regulations need to adapt and allow for printed components. A first step was achieved in *AirMesh*, obtaining a three-year temporary building permit in Singapore. Due to the lack of precedents, the structural calculations were complemented with an empirical test. The completed structure was loaded with an overload of 3 tons with no significant deflection. Deflections were monitored over six months, successively obtaining 3D point clouds from the actual structure with a laser scanner. Figure 10.4 shows the original 3D model superimposed with the scanned information. The results validated the structural assumptions and facilitated the granting of the building permit.

10.4   What Is Next?

The field of 3D Printing in architecture is still at an early stage. Built experimental projects such as those presented in this book demonstrate the potential of this technology in design and construction. Technology is progressing fast, both at the design and manufacturing level. And, as limitations of size, speed, cost, and mechanical performance are disappearing, new opportunities for architects open up. Ongoing research at AirLab is expanding on additive manufacturing in three areas of study.

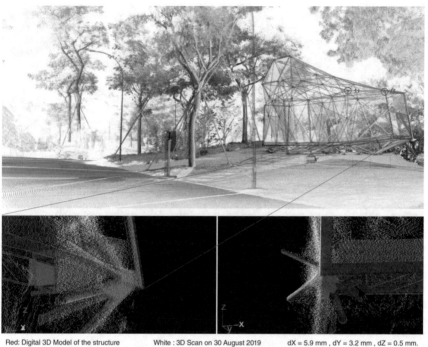

Red: Digital 3D Model of the structure   White : 3D Scan on 30 August 2019   dX = 5.9 mm , dY = 3.2 mm , dZ = 0.5 mm.

Fig. 10.4 Top: 3D cloud with more than 20 millions of collected on-site datapoints, superimposed on the original 3D model. Bottom: Close-up of Node 49 depicting the actual deflection in the structure. The scanning was done with a Faro Focus3D X330 Scanner

10.4.1 AI Design

The formal freedom in design enabled by additive manufacturing means that an immense range of force-driven forms has become technically feasible. To search for new, structurally more efficient designs within this vast design space, artificial intelligence presents a powerful tool. Ongoing design and research work at AirLab explores applications of machine learning algorithms in the design of 3D-printed components. By defining precise structural and boundary conditions, mechanical material properties, and geometrical constraints, multi-objective optimization methods deliver a wide range of design options. The resulting geometric complexity and structural efficiency cannot be produced with conventional design methods, nor manufactured with traditional fabrication tools. The project *AI Table* investigates artificial intelligence in the design of a table. Investment casting with 3D-printed sand molds was identified as a promising technology to produce large-scale, intricate parts. Figure 10.5 shows a functional part of the *AI Table*.

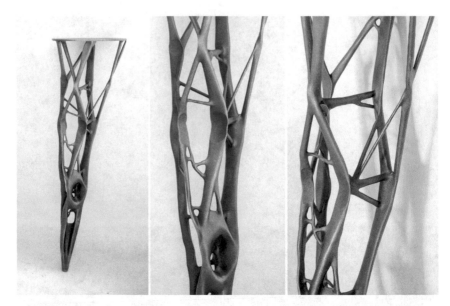

Fig. 10.5 *AI Table* uses artificial intelligence methods to search for table legs design that optimizes the use of material. 75 mm height legs present an organic shape and are cast in bronze using 3D-printed sand molds

10.4.2   Large-Scale Manufacturing

Reliable commercial printers deliver a small printing volume for architectural applications, with typical volumes of below $500 \times 500 \times 500$ mm. Architectural projects can effectively work with these size constraints either by combining printed parts with standard components—see Chap. 3 for an example of this approach—or by large aggregations of printed components—see Chap. 4.

However, at a research level, large-scale additive manufacturing is producing promising results. Typically, these setups include a robotic or gantry manipulator, operating a large volume extruder using inexpensive materials such as concrete, clay, or polymers. With these technologies, large building components can be printed as a single part. AirLab's ongoing work includes a polymer and ceramic projects which implement a robotic setup. Figures 10.6, 10.7 and 10.8 show three ongoing projects using large-scale printing.

10.4.3   Assisted Assembly

The assembly of designs with high geometric complexity represents one of the most challenging aspects, even with well-planned schemes. The large number of unique parts and the complexity of structures makes the identification, positioning, and

Fig. 10.6 A 4 m diameter dome printed with recycled PETG dome. Left: Printing equipment integrated by a robotic arm, bespoke extruder, and mobile printing platform. Center: Rendering of the proposed design for DB Schenker Singapore. Right: Actual prototype of a part printed in recycled PETG

Fig. 10.7 Vertical HDB urban farming project uses robotic ceramic printing

fastening of components a challenge. Recent improvements in augmented reality and virtual reality technologies present great potentials for on-site applications that can support complicated assembly processes.

10.5 Redefining the Role of Connections in Architecture

Many approaches to using 3D Printing in architecture attempt to print everything. This leads to machine limitations, geometric constraints, and high material costs. However, the series of projects presented in this book proposes a different paradigm, combining standard with non-standard materials and one that limits 3D Printing to components that address complex geometrical, structural, and other technical problems.

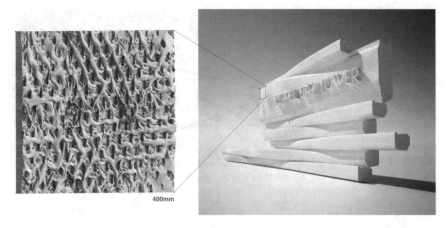

Fig. 10.8 The Keppel Bay Tower signage (right) uses large-scale sand printing as formwork for concrete. The fine resolution of the prints allows for intricate details to be designed and printed, as shown in the close-up view on the left

In this hybrid printed and standard paradigm, rethinking of the role of connections is one of the most relevant topics that will enable the advancement of 3D Printing in architecture. The long tectonic tradition of expressing the flow and transfer of loads through the design of details was de-emphasized by the Modern Movement. For example, the welded steelwork in many of Mies van der Rohe's projects renders the junctions as invisible as possible [1]. The legacy of the Modern Movement crystallized into the standardized connections that are widely used in architectural design today. With 3D Printing, the benefits of standardization no longer exist, and with it, its constraints regarding spatial and structural design. The new tectonics resulting from the printed and standard paradigm constitute a shift toward creating unique transitions which are both structurally logical and aesthetically advantageous [2]. This approach allows for maximum design flexibility and greater performative and ornamental qualities.

The ten projects described in this book propose different strategies for designing connections formally, structurally, and aesthetically, point to line, line to line, and line to surface. This has important implications for the perception of structures as continuum that are composed of multiple standard and non-standard 3D-printed elements.

10.6   Architectural Perspectives

Throughout history, material- and technology-driven changes have had significant impact on architecture. The current speed of development of digital tools and processes is rapidly impacting architectural design through new technologies, including freeform modeling, generative design, simulation of functional performance, and various manufacturing processes. Moving forward, 3D Printing appears to be emerging as an integrating technology that ties all of the above together.

3D Printing can help address major architectural design concerns of our time including environmental sustainability and economic resilience. It offers great potential for new approaches to architecture including designing for better material efficiencies and alternative methods of construction. The formal freedom that 3D Printing allows for presents an opportunity for architects to fundamentally rethink how, what, and why they build. The ideas and projects presented in this book can contribute to this rethinking as well as the larger discourse on contemporary design.

While the projects presented in this volume cover many aspects of 3D Printing and its potential use in architectural design, a broader application of the technology in practice has yet to happen. More research and development work needs to be done, including on how it impacts digital workflows and on its potential for generative design, design optimization, and post-processing, to name just a few. Further areas of exploration include algorithm design, new 3D Printing materials and processes for larger, faster, and more consistent printing, printability analysis for complex 3D-printed geometries, construction legislation, and optimization of supply chains for the manufacture and assembly of large-scale architectural projects. Combined and integrated into increasingly seamless workflows, all of the above will allow for exciting innovations in architectural design.

References

1. Frampton K (1985) Studies in tectonic culture. Harvard University Graduate School of Design, Cambridge
2. Thompson D (1963) On growth and form, 2d edn. Cambridge University Press, Cambridge

Printed in the United States
By Bookmasters